# XVIII International Youth Scientific School on Coherent Optics and Optical Spectroscopy 2014



Journal of Physics: Conference Series Volume 613

## Kazan, Russia
## 27-29 October 2014

ISBN: 978-1-5108-0360-2
ISSN: 1742-6588

**Printed from e-media with permission by:**

Curran Associates, Inc.
57 Morehouse Lane
Red Hook, NY 12571

**Some format issues inherent in the e-media version may also appear in this print version.**



Printed by Curran Associates, Inc. (2015)

For permission requests, please contact the Institute of Physics
at the address below.

Institute of Physics
Dirac House, Temple Back
Bristol BS1 6BE UK

Phone: 44 1 17 929 7481
Fax: 44 1 17 920 0979

techtracking@iop.org

**Additional copies of this publication are available from:**

Curran Associates, Inc.
57 Morehouse Lane
Red Hook, NY 12571 USA
Phone: 845-758-0400
Fax: 845-758-2634
Email: curran@proceedings.com
Web: www.proceedings.com

# TABLE OF CONTENTS

# Preface

Editors: M.Kh.Salakhov, V.V.Samartsev, R.Kh.Gainutdinov

Kazan Federal University has held the annual International Youth School "Coherent Optics and Optical Spectroscopy" since 1997. The choice of the topic is not accidental. Kazan is the home of photon echo which was predicted at Kazan Physical-Technical Institute in 1963 by Prof. U.G. Kopvil'em and V.R. Nagibarov and observed in Columbia University by N.A. Kurnit, I.D. Abella, and S.R. Hartmann in 1964. Since then, photon echo has become a powerful tool of coherent optical spectroscopy and optical information processing, which have been developed here in Kazan in close collaboration between Kazan Physical-Technical Institute and Kazan Federal University. The main subjects of the XVIII International Youth School are: Nonlinear and coherent optics; Atomic and molecular spectroscopy; Coherent laser spectroscopy; Problems of quantum optics; Quantum theory of radiation; and Nanophotonics and Scanning Probe Microscopy. The unchallenged organizers of that school are Kazan Federal University and Kazan E.K. Zavoisky Physical-Technical Institute. The rector of the School is Professor Myakzyum Salakhov, and the vice-rector is Professor Vitaly Samartsev.

The International Youth Scientific School "Coherent Optics and Optical Spectroscopy" follows the global pattern of comprehensive studies of matter properties and their interaction with electromagnetic fields. Since 1997 more than 100 famous scientists from the USA, Germany, Ukraine, Belarus and Russia have given plenary lecture presentations. Here over 1000 young scientists had an opportunity to participate in lively discussions about the latest scientific news. Many young people have submitted interesting reports on photonics, quantum electronics, laser physics, quantum optics, traditional optical and laser spectroscopy, non-linear optics, material science and nanotechnology. Here we are publishing the full-size papers prepared from the most interesting lectures and reports selected by the Program Committee of the School.

The International Youth Scientific School "Coherent Optics and Optical Spectroscopy" was greatly supported by The Optical Society of America, the Russian Foundation for Basic Research, the non-profit Dynasty Foundation, the Tatarstan Academy of Science, and the Ministry of Education and Science of the Russian Federation. It is a pleasure to thank the sponsors and all the participants and contributors who made the International School meeting possible and interesting.

# Peer review statement

All papers published in this volume of *Journal of Physics: Conference Series* have been peer reviewed through processes administered by the proceedings Editors. Reviews were conducted by expert referees to the professional and scientific standards expected of a proceedings journal published by IOP Publishing.

# Influence of the photonic crystal medium on the self-interaction of electrons

**R Kh Gainutdinov[1], M A Khamadeev[1], M R Mohebbifar[1], A S Petrova[1], K A Ziyatdinova[1], M Kh Salakhov[1,2]**

[1] Kazan Federal University, Institute of Physics, 420008, Kremlevskaya St, 18, Kazan, Russian Federation

[2] Tatarstan Academy of Sciences, 420111, Bauman St, 20, Kazan, Russian Federation

E-mail: Renat.Gainutdinov@kpfu.ru

**Abstract.** The self-interaction of an electron in the photonic crystal medium is investigated. We show that modification of the self-interaction of electrons in photonic crystals is very significant and, as a consequence, at some photon frequencies the electron-light interaction becomes effectively strong.

## 1. Introduction

Photonic crystals (PC) are periodic systems that consist of dielectrics with different refractive indices. Its periodic dielectric function yields a photonic band structure with photonic band gaps. It has been shown that such a structure leads to quantum effects, including the coherent control of the spontaneous emission [1], the appearance of photon-atom bound states [2–5], the non-Markovian character of radiative decay [6], enhanced quantum interference effects [7], the localization of superradiance near the photonic band edge [8] and the modification of Lamb shift [2, 3, 9–13]. At the same time, in Ref. [14] it has been shown that there is a novel effect in PC medium which consists in the fact that the rest mass of an electron should change its value. Actually, this effect is another consequence of the periodic changing of the photon-matter interaction in the PC medium. The origin of the effect is the significant modification of the electromagnetic field in the PC medium. In this paper we investigate the influence of the modification of the photonic density of states (DOS) in PC on the self-interaction of electrons.

## 2. Methods

In our investigation we use the method [15] that allows one to solve the problem nonperturbatively. This method is based on the generalized dynamical equation (GDE) that has been derived in [15] as the most general dynamical equation consistent with the current concepts of quantum physics. Being equivalent to the Schrödinger equation in the case when the interaction in a quantum system is instantaneous, GDE allows one to extend dynamics to the case of nonlocal-in-time interactions. This equation provides a new insight into many problems in atomic physics [17–20], nuclear physics [21–24] and quantum optics [25, 26]. The method allows one to take into account from the every beginning that the contribution to the Green operator $G(z)$, which comes from the processes associated with the self-interaction of particles, has the same structure as the free Green operator $G_0(z)$. For this reason it is natural to replace $G_0(z)$ by the operator $G_0^{(v)}(z)$, which describes the evolution of the system when particles propagate freely or interact with vacuum, and, hence, has the structure

$$\langle n'|\, G_0^{(v)}(z)\, |n\rangle = \langle n' \mid n\rangle \left(z - E_n - C_n(z)\right)^{-1},$$

with $|n\rangle$ being the eigenvectors of the free Hamiltonian ($H_0\,|n\rangle = E_n\,|n\rangle$). Other contributions are described by the operator $G^{(I)}(z) = G_0^{(v)}(z)M(z)G_0^{(v)}(z)$:

$$G(z) = G_0^{(v)}(z) + G^{(I)}(z) \equiv G_0^{(v)}(z) + G_0^{(v)}(z)M(z)G_0^{(v)}(z),$$

where the operator $M(z)$ describes the processes in which some particles interact each with other. The equations for $C(z)$ and $M(z)$ are derived from GDE. The equation for the function $C_n(z)$ referred to as the self-energy function reads

$$\frac{dC_n(z)}{dz} = -\,\langle n|\, M(z) \left(G_0^{(v)}(z)\right)^2 M(z)\, |n\rangle\,,\ \langle n \mid n\rangle = 1,$$

and the condition

$$z - E_n^{(0)} - C_n(z) = 0 \tag{1}$$

determines the physical masses of particles. In the case when we deal with an atom and $|n\rangle$ describes an atomic state, equation (1) determines the self-energy correction (the Lamb shift) to the energy $E_n$ of the state $|n\rangle$. An approximative solution of this equation is $E_n \equiv E_n^{(0)} + C_n^{(0)}(E_n^{(0)})$. For this approximation to be valid the variation of $C_n(z)$ in the energy interval between $E_n^{(0)}$ and $E_n$ must be negligible. This is the case for atoms in free space. In fact at leading order in $\alpha$ the equation for $C_n(z)$ is reduced to the equation [27, 28]

$$\frac{dC_n^{(0)}(z)}{dz} = -\,\langle n|\, H_I \left(G_0^{(v)}(z)\right)^2 H_I\, |n\rangle\,,\ \langle n \mid n\rangle = 1, \tag{2}$$

with $H_I$ being the interaction Hamiltonian. By solving this equation with an appropriate boundary condition we arrive at the ordinary expressions for the self-energy shifts and widths of energy levels. However, in the case of atoms in the PC medium the variation of the self-energy function in the relevant vicinity of the point $z = E_n^{(0)}$ can be very significant and, as a result, the above approximation is invalid. In this case one has to derive the self-energy function from a nonperturbative solution of the equations for $C_n(z)$ and $M(z)$.

## 3. Photon density of states and the electron self-energy function in the PC medium

Let us consider the self-interaction of an electron in a PC. The Bloch structure of the photon states in the PC medium means that these states can be expanded in a set of Bloch states. The interaction Hamiltonian describing the light-matter coupling in the PC medium is $H_I^{pc} = -\frac{e}{m_e}\mathbf{p}\cdot\mathbf{A}_{pc}$, where $\mathbf{p}$ is the electron momentum, and $\mathbf{A}_{pc}$ is the PC-medium quantized vector potential (for details see [14]). Let us assume that the perturbation theory is applicable in this case, and the leading order solution for $M(z)$ is $M(z) = H_I^{pc}$. Equation (2) for the self-energy function is reduced in this case to the equation (here and below we use the unit system where $\hbar = c = \varepsilon_0 = 1$)

$$\frac{dC_{\mathbf{p}}^{pc}(z)}{dz} = -\sum_{\mathbf{p}'}\sum_{\mathbf{k},\varepsilon_\lambda} \frac{\langle \mathbf{p}|H_I^{pc}|\mathbf{p}';\mathbf{k},n\rangle\langle \mathbf{p}';\mathbf{k},n|H_I^{pc}|\mathbf{p}\rangle}{\left(z-m_e-\frac{\mathbf{p}'^2}{2m_e}-\omega_{\mathbf{k}n}\right)^2} + \sum_{\mathbf{p}'}\sum_{\mathbf{k},n} \frac{\langle \mathbf{p}|H_I|\mathbf{p}';\mathbf{k},\varepsilon_\lambda\rangle\langle \mathbf{p}';\mathbf{k},\varepsilon_\lambda|H_I|\mathbf{p}\rangle}{\left(z-m_e-\frac{\mathbf{p}'^2}{2m_e}-|\mathbf{k}|\right)^2}, \tag{3}$$

where $|\mathbf{k},n\rangle$ is the photon in Bloch state with energy $\omega_{\mathbf{k}n}$. Actually, $C^{pc}_{\mathbf{p}}(z)$ is the difference between the self-energy function in a PC and that in vacuum. The second term of the right-hand part of equation (3) is $dC^{(v)}_{\mathbf{p}}(z)/dz$ where $C^{(v)}_{\mathbf{p}}(z)$ is the self-energy function of the electron in vacuum. The solution of equation (3) is

$$C^{pc}_{\mathbf{p}}(z) = \sum_{\mathbf{p}'}\sum_{\mathbf{k},\varepsilon_\lambda} \frac{\langle \mathbf{p}|H^{pc}_I|\mathbf{p}';\mathbf{k},n\rangle\langle \mathbf{p}';\mathbf{k},n|H^{pc}_I|\mathbf{p}\rangle}{z-m_e-\frac{\mathbf{p}'^2}{2m_e}-\omega_{\mathbf{k}n}} - \sum_{\mathbf{p}'}\sum_{\mathbf{k},n} \frac{\langle \mathbf{p}|H_I|\mathbf{p}';\mathbf{k},\varepsilon_\lambda\rangle\langle \mathbf{p}';\mathbf{k},\varepsilon_\lambda|H_I|\mathbf{p}\rangle}{z-m_e-\frac{\mathbf{p}'^2}{2m_e}-|\mathbf{k}|}$$

The value of $C^{pc}_{\mathbf{p}}(z)$ at $z = m_e + E_{\mathbf{p}}$ is $C^{pc}_{\mathbf{p}}(z = m_e + E_{\mathbf{p}}) = \left(1 - \frac{\mathbf{p}^2}{2m_e^2}\right)\delta m_{pc}$, with

$$\delta m_{pc} = -\frac{2m_e^2}{\mathbf{p}^2}\left(\sum_{\mathbf{p}'}\sum_{\mathbf{k},\varepsilon_\lambda} \frac{\langle \mathbf{p}|H^{pc}_I|\mathbf{p}';\mathbf{k},n\rangle\langle \mathbf{p}';\mathbf{k},n|H^{pc}_I|\mathbf{p}\rangle}{\frac{\mathbf{p}^2}{2m_e}-\frac{\mathbf{p}'^2}{2m_e}-\omega_{\mathbf{k}n}} - \sum_{\mathbf{p}'}\sum_{\mathbf{k},n} \frac{\langle \mathbf{p}|H_I|\mathbf{p}';\mathbf{k},\varepsilon_\lambda\rangle\langle \mathbf{p}';\mathbf{k},\varepsilon_\lambda|H_I|\mathbf{p}\rangle}{\frac{\mathbf{p}^2}{2m_e}-\frac{\mathbf{p}'^2}{2m_e}-|\mathbf{k}|}\right) \quad (4)$$

being the PC medium correction to the electron mass derived in [14]. Expression (4) for $\delta m_{pc}$ can be simplified [14]:

$$\delta m_{pc} = \frac{\alpha}{\pi^2}\int d\omega \frac{N_{\mathbf{p}}(\omega) - \omega^2}{\omega^2},$$

where $N_{\mathbf{p}}(\omega) = N_{DOS}(\omega)D_{\mathbf{p}}(\omega)$ with $N_{DOS}$ being the photon density of states and $D_{\mathbf{p}}(\omega) = \sum_{\mathbf{G},\lambda}\left|e^{\mathbf{G}\lambda}_{\mathbf{k}n}\right|^2 (\varepsilon_{\mathbf{G}\lambda})^2_{\mathbf{p}}\Big|_{\omega_{\mathbf{k}n}=\omega}$. Here $e^{\mathbf{G}\lambda}_{\mathbf{k}n}$ are the amplitudes of the components, which constitute the coefficients of the plane wave expansion of the Bloch wave function of the photon, $(\varepsilon_{\mathbf{G}\lambda})_p$ are the polarization vector projections onto the electron momentum axis. In the similar way equation (3) can be rewritten in the form

$$\frac{dC^{pc}_e(z)}{dz} = -\frac{\alpha}{\pi^2}\int \frac{d\omega}{\omega}\frac{N_{\mathbf{p}}(\omega) - \omega^2}{(z - m_e - E_{\mathbf{p}} - \omega)^2}.$$

The DOS describes the number of states per interval of energy at each energy level that are available to be occuped, and in vacuum is equal to $\omega^2$. The modification of the electromagnetic field in the PC medium results in the fact that the DOS in the PC medium differs profoundly from that in vacuum. This in turn has a significant effect on the self-energy function of the electron. Further estimates can be made with the model of $N_{\mathbf{p}}(\omega)$ analogous to the one proposed in [14]. The results of our calculations of the self-energy function $C_{\mathbf{p}}(z)$ displayed in figure 1 show its strong dependence on $z$. In the same way as it was made in Ref. [14] we can derive the formula for the self-energy function of the atomic hydrogen in the state $|N\rangle = |n,j,l,m\rangle$

$$C_N(z) = \int d^3p \Psi^*_N(\mathbf{p})C^{pc}_{\mathbf{p}}(z)\Psi_N(\mathbf{p}),$$

where $\Psi_N(\mathbf{p})$ is the wave function in momentum space. Obviously, the above strong variation of the self-energy functions of electrons in the PC medium should result in a strong modification of the atomic self-energy function. In the case of atoms the self-energy function describes also the decay law of unstable states. The standard treatment of unstable states is based on the assumption that they correspond to poles of the analytic continuation of the Green operator. The positions of these poles are determined by equation (1). The approximative correction to the energy level determined by this equation is $C^{(0)}_n(E^{(0)}_n) = \Delta E_n - \frac{i}{2}\Gamma_n$, with $\Delta E_n$ and $\Gamma_n$ being the self-energy shift and the natural width of the energy level of the state $|n\rangle$ respectively, and the Green operator $G^{(v)}_0(z)$ takes the form

$$G^{(v)}_0(z) = \sum_n \frac{|n\rangle\langle n|}{z - E_n - \Delta E_n - \frac{i}{2}\Gamma_n}, \quad (5)$$

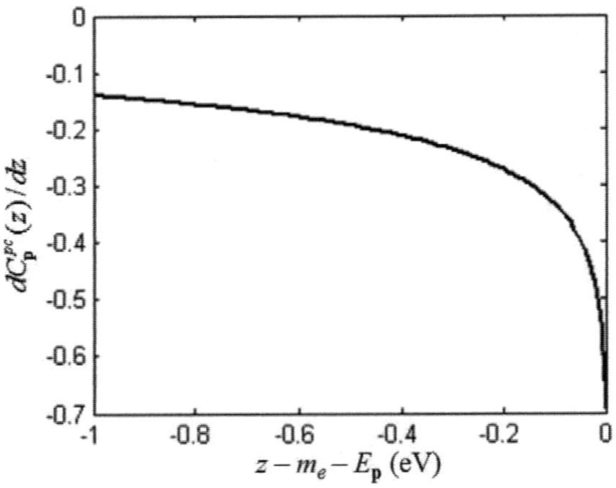


**Figure 1.** The self-energy function $C_{\mathbf{p}}^{pc}(z)$ of a free electron placed in PC medium. The calculations was performed with the model of $N_{\mathbf{p}}(\omega)$ analogous to the one proposed in [14] with the parameters $n_{eff} = 4$, $\mu = 15$ eV, $\tau = 0.01$ eV, $\omega_0 = 1$ eV, $h_0 = 0.96$, $\sigma_0 = 0.07$ eV

which implies that the decay law is exponential and hence $\Gamma_n$ is equal to the spontaneous rate $\gamma_n$. For this quasi-stationary approximation to be valid the variation of the self-energy function should be negligible. However, as we have shown this is not the case for the self-energy functions of electrons and atoms in the PC medium. This means that the decay law of the unstable states of atoms placed in the PC medium can be nonexponential. Note in this connection that as it follows from the Fock-Krylov theorem, in general an unstable atomic state should be described by an energy distribution. Only in the case in which the variation in $z$ of the self-energy function near the position of the pole corresponding to the unstable state is negligible the energy distribution is of the Breit-Wigner shape and hence the state can be associated with some complex energy. However, as we have seen this is not the case when we deal with atoms in the PC medium. As no discrete energy level can be associated with the unstable state in this case it is not an eigenstate of any Hamiltonian. Therefore the methods which are based on the eigenstates of any Hamiltonian fail in this case. At the same time, the approach based on the generalized dynamical equation allows one to regard an unstable state as an energy distribution from the very beginning. The definition of the Green operator in generalized quantum dynamics is more general and remains valid even for the systems, where no Hamiltonian can be found as an operator generating dynamics of the system. From the results of calculations presented in figure 1 it follows that the differential of the self-energy function in the PC medium is $10^3$ times larger in magnitude than that in vacuum. On the other hand, as it follows from the equation for operator $M(z)$, there is a direct connection between the value of the differential of the leading-order vertex and the intensity of the light-matter interaction. This means that at some photon frequencies the electron-light interaction becomes effectively strong and equations for the self-energy function and the operator $M(z)$ should be solved nonperturbatively.

The above results show that the strong modification of the electromagnetic field in the PC medium gives rise to the fact that the electron self-energy function in a PC can significantly differ from that in vacuum. This modification of the electromagnetic field should also have effect on the light-matter interaction in the PC medium. In some cases this interaction can be efficiently strong. It is important because achieving the strong coupling regime is one of the ways of realizing a quantum interface between light and matter [29–32]. In the strong coupling regime the equation for $C(z)$ and $M(z)$ must be solved nonperturbatively. The above way of achieving the strong coupling regime can be improved by using quantum dots as emitters in PC's instead of atoms. In fact, transition dipole moments of energy levels of quantum dots can be much larger than that of atoms and these dipole moments themselves can be significantly increased by the modification of the electromagnetic field in the PC medium.

## 4. Conclusion

We have shown that the self-energy functions of electrons and atoms in the PC medium strongly variates with $z$. This has a significant effect on the character of their interaction with the electromagnetic field. Energy distributions of unstable states of atoms in PC's can differ dramatically from the Breit-Wigner form. This in turn results in the fact that their decay law is in general nonexponential. The significant modification of the self-energy functions in the PC medium results also in the fact that the interaction of atoms with some modes of the electromagnetic field can be effectively strong.

## Acknowledgments

This work was funded by the subsidy allocated to Kazan Federal University for the state assignment in the sphere of scientific activities.

## References

[1] Quang T, Woldeyohannes M, John S and Agarwal G S 1997 *Phys. Rev. Lett.* **79** 5238
[2] John S and Wang J 1990 *Phys. Rev. Lett.* **64** 2418
[3] John S and Wang J 1991 *Phys. Rev.* B **43** 12772
[4] Bay S, Lambropoulos P and Mölmer K 1997 *Phys. Rev. Lett.* **79** 2654
[5] Bay S, Lambropoulos P and Mölmer K 1997 *Phys. Rev.* A **55** 1485
[6] Busch K, Vats N, John S and Sanders B C 2000 *Phys. Rev.* E **62** 4251
[7] Zhu S Y, Chen H and Huang H 1997 *Phys. Rev. Lett.* **79** 205
[8] John S and Quang T 1995 *Phys. Rev. Lett.* **74** 3419
[9] Zhu S Y, Yang Y, Chen H, Zheng H and Zubairy M S 2000 *Phys. Rev. Lett.* **84** 2136
[10] Vats N, John S and Busch K 2002 *Phys. Rev.* A **65** 043808
[11] Li Z Y and Xia Y 2001 *Phys. Rev.* B **63** 121305(R)
[12] Wang X-H, Kivshar Y S and Gu B-Y 2004 *Phys. Rev. Lett.* **93** 073901
[13] Wang X-H, Kivshar Y S and Gu B-Y 2005 *Science and Technology of Advanced Materials* **6** 814
[14] Gainutdinov R Kh, Khamadeev M A and Salakhov M Kh 2012 *Phys. Rev.* A **85** 053836
[15] Gainutdinov R Kh 1999 *J. Phys.* A **32** 5657
[16] Gainutdinov R Kh, Khamadeev M A and Salakhov M Kh 2013 *J. Phys.: Conf. Ser.* **478** 012017
[17] Gainutdinov R Kh, Mutygullina A A and Scheid W 2001 *Proc. of SPIE* **4605** 214
[18] Gainutdinov R Kh, Mutygullina A A and Scheid W 2001 *Proc. of SPIE* **4605** 222
[19] Gainutdinov R Kh 1995 *ZETP* **108** 1600
[20] Gainutdinov R Kh 1986 *Optika i Spektroskopiya* **60** 890
[21] Gainutdinov R Kh and Moutygoullina A A 1997 *Physics of Atomic Nuclei* **60** 841
[22] Gainutdinov R Kh and Moutygoullina A A 1999 *Physics of Atomic Nuclei* **62** 1905
[23] Gainutdinov R Kh and Mutygullina A A 2002 *Physics of Atomic Nuclei* **65** 1421
[24] Gainutdinov R Kh, Mutygullina A A, Khamadeev M A and Petrova A S 2013 *J. Phys.: Conf. Ser.* **478** 012019
[25] Gainutdinov R Kh and Mutygullina A A 2013 *J. Phys.: Conf. Ser.* **478** 012018
[26] Gainutdinov R Kh and Mutygullina A A 2009 *Physica Scripta* **T135** 014021
[27] Gainutdinov R Kh 1984 *Ukrainskii fizicheskii zhurnal* **29** 805
[28] Gainutdinov R Kh 1989 *J. Phys.* A **22** 269
[29] Kimble H J 2008 *Nature* **453** 1023
[30] Pellizzari T, Gardiner S A, Cirac J I and Zoller P 1995 *Phys. Rev. Lett.* **75** 3788
[31] Walls D F and Milburn G J 2008 *Quantum Optics* (Berlin: Springer-Verlag) p 425
[32] Kimble H J 1998 *Phys. Scr.* **T76** 127

Journal of Physics: Conference Series **613** (2015) 012002 doi:10.1088/1742-6596/613/1/012002

# Tunneling-induced self-energy shift of energy levels of a quantum dot

**R Kh Gainutdinov, M A Khamadeev, M R Mohebbifar and A A Mutygullina**

Optics and Nanophotonics Department, Institute of Physics, Kazan Federal University, 18 Kremlevskaya St, Kazan 420008, Russia

E-mail: `Renat.Gainutdinov@kpfu.ru`

**Abstract.**
The self-interaction of a quantum dot tunnel with Coulomb interaction coupled to two noninteracting leads is investigated. The self-energy function describing this interaction is added to a bare energy of a dot state. In the standard way of determining the self-interaction (tunneling-induced) corrections to bare energies of emitters (atoms, quantum dots, etc) the variations of the self-energy functions with energy are ignored, and these corrections are considered to be equal to the values of the self-energy functions for bare energies of states. We show that actually in the case of quantum dots the variations of the self-energy functions in the energy interval between the bare and true energies can be strong, and this can have a significant effect on the values of the tunneling-induced shifts of energy levels of quantum dots.

## 1. Introduction

Semiconductor quantum dots are nanometer-sized three-dimensional structures which confine electrons in all three directions. This confinement gives rise to a discrete energy spectrum [1]. Quantum dots have offered unique opportunities to investigate quantum optical effects in solid state systems such as quantum interference [2], Rabi oscillations [3–7], photon anti-bunching [8], etc. The remarkable feature of quantum dots is that the interaction of electromagnetic field with them can differ profoundly from that with ordinary atoms. This fact manifests itself in the recently observed [9] Mollow triplet from self-assembled quantum dots. The observed resonance fluorescence spectrum showed a dramatical disagreement with the predictions of quantum optics [10]. This disagreement can originate from the fact that in the standard theory of the resonance fluorescence from a two-level atom the nondecay transitions caused by the self-interaction processes are not taken into account. In the case of quantum dots the role of the virtual photons in the self-interaction of ordinary atoms can be played, for example, by electrons that leave the quantum dot and then come back. Such a self-interaction can be much more significant than the interaction of an atom with its own radiation field. In this paper we investigate the effect of such a self-interaction on the tunneling-induced shift of energy levels of a quantum dot.

## 2. Methods

In our investigation we use the method [11] that allows one to solve the problem nonperturbatively. This method is based on the generalized dynamical equation (GDE) that has been derived in [11] as the most general dynamical equation consistent with the current concepts of quantum physics. Being equivalent to the Schrödinger equation in the case when the interaction in a quantum system is instantaneous, GDE allows one to extend dynamics to the case of nonlocal-in-time interactions. This equation provides a new insight into many problems in atomic physics [12–15], nuclear physics [16–19] and quantum optics [20, 21]. The method allows one to take into account from the every beginning that the contribution to the Green operator $G(z)$, which comes from the processes associated with the self-interaction of particles, has the same structure as the free Green operator $G_0(z)$. For this reason it is natural to replace $G_0(z)$ by the operator $G_0^{(v)}(z)$, which describes the evolution of the system when particles propagate freely or interact with vacuum, and, hence, has the structure

$$\langle m'|\, G_0^{(v)}(z)\, |m\rangle = \langle m' \mid m\rangle\, (z - E_m - C_m(z))^{-1}, \tag{1}$$

with $|m\rangle$ being the eigenvectors of the free Hamiltonian ($H_0\,|m\rangle = E_m\,|m\rangle$). Other contributions are described by the operator $G^{(I)}(z) = G_0^{(v)}(z)M(z)G_0^{(v)}(z)$:

$$G(z) = G_0^{(v)}(z) + G^{(I)}(z) \equiv G_0^{(v)}(z) + G_0^{(v)}(z)M(z)G_0^{(v)}(z),$$

where the operator $M(z)$ describes the processes in which some particles interact each with other. The equations for $C(z)$ and $M(z)$ are derived from GDE. The equation for the function $C_m(z)$ referred to as the self-energy function reads

$$\frac{dC_m(z)}{dz} = -\,\langle m|\, M(z) \left(G_0^{(v)}(z)\right)^2 M(z)\, |m\rangle\,,\ \langle m \mid m\rangle = 1, \tag{2}$$

and the condition

$$z - E_m^{(0)} - C_m(z) = 0 \tag{3}$$

determines the physical masses of particles. In the case when we deal with an atom and $|m\rangle$ describes an atomic state, equation (3) determines the self-energy correction (the Lamb shift) to the energy $E_m$ of the state $|m\rangle$. An approximative solution of this equation is $E_m \equiv E_m^{(0)} + C_m^{(0)}(E_m^{(0)}) \equiv E_m^{(0)} + \Delta E_m^L - \frac{i}{2}\Gamma_m$, with $\Delta E_m^L$ and $\Gamma_m$ being the Lamb shift and the natural width of the energy level of the state $|m\rangle$ respectively. For this approximation to be valid the variation of $C_m(z)$ in the energy interval between $E_m^{(0)}$ and $E_m$ must be negligible. This is the case for atoms in free space. In fact, at leading order in $\alpha$ the equation for $C_m(z)$ is reduced to the equation [22, 23]

$$\frac{dC_m^{(0)}(z)}{dz} = -\,\langle m|\, H_I \left(G_0^{(v)}(z)\right)^2 H_I\, |m\rangle\,,\ \langle m \mid m\rangle = 1, \tag{4}$$

with $H_I$ being the interaction Hamiltonian. By solving this equation with an appropriate boundary condition we arrive at the ordinary expressions for the self-energy shifts and widths of energy levels. However, in the case of quantum dots the variation of the self-energy function in the relevant vicinity of the point $z = E_m^{(0)}$ can be very significant and, as a result, the above approximation is invalid. In this case the self-interaction function can not be parameterized by a shift and a width, and one has to derive the self-energy function from a nonperturbative solution of the equations for $C_m(z)$ and $M(z)$.

## 3. The self-energy functions of the quantum dot states

There are several different kinds of quantum dots which differ in both their method of manufacture and their electronic and optical properties. In this work we investigate the self-interaction of a single-level quantum dot with arbitrary strong one-site Coulomb interaction tunnel coupled to two noninteracting leads. This model was used in [24] for analyzing tunneling-induced quantum fluctuation in such quantum dots. The Hamiltonian of system consisting of the quantum dot and leads is described by $H = H_{qd} + H_{tun} + H_{lead}$, where $H_{qd}$ is the Hamiltonian of the quantum dot, $H_{tun}$ is the Hamiltonian of tunneling process between the leads and the quantum dot and $H_{lead}$ is the Hamiltonian of the leads. The quantum dot can be described by the single-level Anderson impurity model, $H_{qd} = \sum_{\sigma=\uparrow,\downarrow} \varepsilon d_\sigma^\dagger d_\sigma + U n_\uparrow n_\downarrow$, where the creation (annihilation) operator for an electron with spin $\sigma$ on the dot is given by $d_\sigma^\dagger$ and $d_\sigma$, $n_\sigma = d_\sigma^\dagger d_\sigma$ is the corresponding number operator. The on-site repulsion $U$ describes the energy cost for double occupation (when we have two electrons in the quantum dot). The Hamiltonian $H_{tun}$ is given by $H_{tun} = \sum_{\alpha,k,\sigma} V_\alpha c_{\alpha,k,\sigma}^\dagger d_\sigma + H.c.$, where $V_\alpha$ is the momentum and spin-independent tunnel matrix element, $c_{\alpha,k,\sigma}^\dagger (c_{\alpha,k,\sigma})$ is the creation (annihilation) operator for electrons with spin $\sigma$ and momentum $k$ in lead $\alpha = l, r$. The Hamiltonian $H_{lead}$ is given by $H_{lead} = \sum_{\alpha,k,\sigma} \varepsilon_{\alpha,k} c_{\alpha,k,\sigma}^\dagger c_{\alpha,k,\sigma}$. The chemical potentials of the two leads differ by the applied bias $\mu_l - \mu_r = -eV$. We assume that the density of states $\rho_\alpha$ in the leads is constant for transport and defines the tunnel coupling strength $R_\alpha$ as $R_\alpha = 2\pi\rho_\alpha \mid V_\alpha \mid^2$, where $V_\alpha$ is the tunnel matrix element and $R = R_l + R_r$. We will denote the states of quantum dots as $|\Psi_0\rangle$ for an empty dot, $|\Psi_\sigma\rangle$ for a singly occupied dot with spin $\sigma =\uparrow,\downarrow$ and $|\Psi_d\rangle$ for a doubly occupied dot. In this model we assume that reservoirs are in equilibrium and we average over the reservoir part of the initial states according to the Fermi distribution

$$f_\alpha(E) = \frac{1}{1 + \exp(\frac{E-\mu_\alpha}{K_B T})}.$$

Solving the leading-order equation (4) yields the following expressions for the self-energy functions $C_0(z)$, $C_\sigma(z)$ and $C_d(z)$ of the quantum-dot states $|\Psi_0\rangle$, $|\Psi_\sigma\rangle$ and $|\Psi_d\rangle$ respectively:

$$C_0(z) = 2\sum_\alpha \frac{R_\alpha}{2\pi} \int d\omega \frac{f_\alpha(\omega)}{z - \varepsilon + \omega} \tag{5}$$

$$C_\sigma(z) = \sum_\alpha \frac{R_\alpha}{2\pi} \int d\omega \left( \frac{1 - f_\alpha(\omega)}{z - \omega} + \frac{f_\alpha(\omega)}{z - 2\varepsilon - U + \omega} \right) \tag{6}$$

$$C_d(z) = 2\sum_\alpha \frac{R_\alpha}{2\pi} \int d\omega \frac{1 - f_\alpha(\omega)}{z - \varepsilon - \omega} \tag{7}$$

Let us assume that the variations of these self-energy functions with $z$ are weak. In this case $\Delta E_0^{(ap)} = C_0(E_0)$, $\Delta E_\sigma^{(ap)} = C_\sigma(E_\sigma)$ and $\Delta E_d^{(ap)} = C_d(E_d)$ can be considered as the approximative energy shifts of the corresponding energy levels of the quantum dot. The imaginary parts of these shifts $i\text{Im}E_0^{(ap)}$, $i\text{Im}E_\sigma^{(ap)}$ and $i\text{Im}E_d^{(ap)}$ describe dissipative phenomena. Note that the expressions for the energy shifts that follow from equations (5)-(7), are just the same that have been derived in Ref. [24] in second order in the tunnel matrix element $V_{n'n}$. However, as it follows from the results of our calculations the variations with $z$ of the self-energy functions of the quantum-dot states are strong, and one has to solve equation (3) exactly. Figure 1 where the results of calculations of the self-energy correction

$$\delta\varepsilon = \Delta E_\sigma - \Delta E_0 \tag{8}$$

are depicted, shows that the energy shifts obtained in this way can differ dramatically from their approximative values. In our calculations we have used the fact that the part that is independent of the Fermi distribution function of $C_\sigma(z)$ given by equation (6) can be included

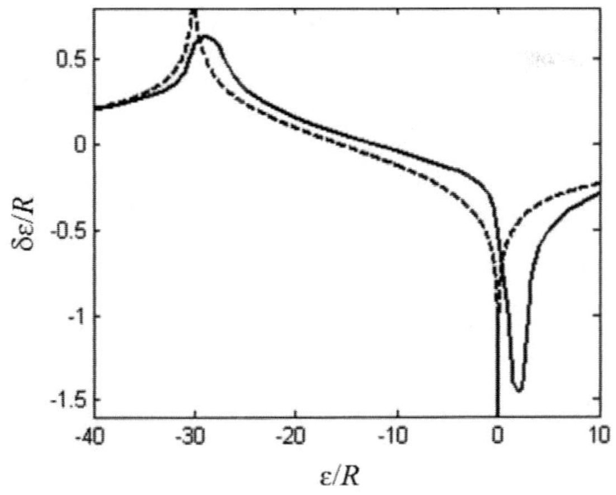


**Figure 1.** Results of calculation of the real part of the corrections $\delta\varepsilon$ defined by equation (9) (solid line) and equation (10) (dashed line) for the parameters $U = 30R$, $\mu = 300R$ at the zero temperature and in the assumption that the left lead is the same as the right lead.

into the correction to the energy of empty level. In this way from equations (5), (6) and (8) we get

$$\delta\varepsilon = \sum_\alpha \frac{R_\alpha}{2\pi}\int d\omega \left(\frac{-f_\alpha(\omega)}{z_\sigma - \omega} + \frac{f_\alpha(\omega)}{z_\sigma - 2\varepsilon - U + \omega} - 2\frac{f_\alpha(\omega)}{z_0 - \varepsilon + \omega}\right), \tag{9}$$

where $z_0$ and $z_\sigma$ are the solutions of equation (3) both for empty and for single occupation energy levels. If we do not take into account the variation of the self-energy function, and put in equations (5) and (6) $z = 0$ and $z = \varepsilon$ respectively we arrive at the expression derived in [24]:

$$\delta\varepsilon^{(ap)} = -\sum_\alpha \frac{R_\alpha}{2\pi}\int d\omega \left(\frac{f_\alpha(\omega)}{\omega - \varepsilon} + \frac{f_\alpha(\omega)}{\varepsilon + U - \omega}\right). \tag{10}$$

## 4. Conclusions

Starting with equation (4) we have derived expressions (5), (6) and (7) for the self-energy functions of the quantum-dot states. Setting $z = E_0$, $z = E_\sigma$ and $z = E_d$ in equation (5), equation (6) and equation (7) respectively we arrive at the expressions for the energy shifts in quantum dots derived in Ref. [24]. This is the manifestation of the fact that in the standard way of solving the problem the energy shift is assumed to be equal to the value of the real part of the corresponding self-energy function at the bare energy. The above approximative solution of equation (2) gives rise just to the same result. If the variations of the self-energy functions $C_0(z)$, $C_\sigma(z)$ and $C_d(z)$ were weak enough, then in solving equation (2) one could restrict oneself to this approximative solution. However, as it follows from the results of calculation, this is not the case, and equation (3) must be solved exactly. Figure 1 shows that the self-interaction energy shifts obtained by solving equation (3) exactly differ dramatically from the approximative shifts $\mathrm{Re}C_m(E_m)$.

**Acknowledgments**
This work was funded by the subsidy allocated to Kazan Federal University for the state assignment in the sphere of scientific activities.

Journal of Physics: Conference Series **613** (2015) 012002 doi:10.1088/1742-6596/613/1/012002

## References

[1] Michler P 2003 *Single Quantum Dots Fundamentals, Applications and New Concepts* (Berlin: Springer)
[2] Bonadeo N H, Erland J, Gammon D, Park D, Katzer D S and Steel D G 1998 *Science* **282** 1473
[3] Htoon H, Takagahara T, Kulik D, Baklenov O, Holmes A L and Shih C K 2002 *Phys. Rev. Lett.* **88** 87401
[4] Kamada H, Gotoh H, Temmyo J, Takagahara T and Ando H 2001 *Phys. Rev. Lett.* **87** 246401
[5] Patton B, Woggon U and Langbein W 2005 *Phys. Rev. Lett.* **95** 266401
[6] Stievater T H, Li X, Steel D G, Gammon D, Katzer D S, Park D, Piermarocchi C and Sham L J 2001 *Phys. Rev. Lett.* **87** 133603
[7] Zrenner A, Beham E, Stufler S, Findeis F, Bichler M and Abstreiter G 2002 *Nature* **418** 612
[8] Michler P, Kiraz A, Becher C, Schoenfeld W V, Petroff P M, Zhang L, Hu E and Imamoglu A 2000 *Science* **290** 2282
[9] Vamivakas A N, Zhao Y, Lu C-Y and Atatüre M 2009 *Nature Phys.* **327** 840
[10] Scully M and Zubairy M 1997 *Quantum Optics* (Cambridge: Cambridge University Press)
[11] Gainutdinov R Kh 1999 *J. Phys.* A **32** 5657
[12] Gainutdinov R Kh, Mutygullina A A and Scheid W 2001 *Proc. of SPIE* **4605** 214
[13] Gainutdinov R Kh, Mutygullina A A and Scheid W 2001 *Proc. of SPIE* **4605** 222
[14] Gainutdinov R Kh 1995 *ZETP* **108** 1600
[15] Gainutdinov R Kh 1986 *Optika i Spektroskopiya* **60** 890
[16] Gainutdinov R Kh and Moutygoullina A A 1997 *Physics of Atomic Nuclei* **60** 841
[17] Gainutdinov R Kh and Moutygoullina A A 1999 *Physics of Atomic Nuclei* **62** 1905
[18] Gainutdinov R Kh and Mutygullina A A 2002 *Physics of Atomic Nuclei* **65** 1421
[19] Gainutdinov R Kh, Mutygullina A A, Khamadeev M A and Petrova A S 2013 *J. Phys.: Conf. Ser.* **478** 012019
[20] Gainutdinov R Kh and Mutygullina A A 2013 *J. Phys.: Conf. Ser.* **478** 012018
[21] Gainutdinov R Kh and Mutygullina A A 2009 *Physica Scripta* **T135** 014021
[22] Gainutdinov R Kh 1984 *Ukrainskii fizicheskii zhurnal* **29** 805
[23] Gainutdinov R Kh 1989 *J. Phys.* A **22** 269
[24] Splettstoesser J, Governale M and König J 2012 *Phys. Rev.* B **86** 035432

Journal of Physics: Conference Series **613** (2015) 012003 doi:10.1088/1742-6596/613/1/012003

# Natural spectral-line broadening in atoms with unstable nuclei

**R Kh Gainutdinov, A A Mutygullina, A S Petrova**

Kazan Federal University, 18 Kremlevskaja St., Kazan 420008, Russia

E-mail: Renat.Gainutdinov@kpfu.ru

**Abstract.** Spontaneous emission from atoms with unstable nuclei is investigated. The decay law of an unstable nucleus is shown to be not exponential and to depend on a parameter characterizing the self-interaction of the unstable nucleus. This parameter, as well as the decay width, can be extracted from the line profile. It is demonstrated on the instance of Beryllium atom with the unstable neutron-rich $^{13}$Be nucleus that for light atoms this parameter may have an essential effect on their spontaneous emission spectrum. Spontaneous emission from supercritical atoms is investigated as well. Their spectrum is continuous and depends on the decay width of the supercritical nucleus. The most sensitivity to the magnitude of the decay width is observed in the X-ray band.

## 1. Introduction

Atomic spectra are known to be a rich source of information about the structure of atomic nuclei. The interaction between the nucleus and atomic electrons is purely electromagnetic, its character is well known, that is why nuclear parameters can be extracted from atomic spectra without involving any model concepts. In Ref. [1] it was shown that spectroscopic methods may appear very promising for the investigation of unstable nuclei. The instability of the atomic nucleus has an unavoidable effect on the process of spontaneous emission, and hence, the spectrum carries information about the decay parameters of the unstable nucleus. The method of spectroscopic analysis developed in [1] is based on the approach of generalized quantum dynamics (GQD) [2]. Its advantage is that it allows one to regard an unstable system as an energy distribution from the very beginning and does not appeal to the eigenvalues of any Hamiltonian. Using this approach, it was shown [1] that the spectrum of a supercritical atom is continuous and depends not only on the decay width and the mass excess of the superheavy nucleus, but on a parameter characterizing the self-interaction of the nucleus. The developed approach is universal and may be applied to any unstable nucleus. In the present paper we proceed with the investigating the spontaneous emission spectrum from supercritical atoms and search for the spectral band especially sensitive to the magnitude of the decay width of the supercritical nucleus. Besides, we show how the above approach can be applied to the exploration of light unstable nuclei. We regard the spontaneous emission from the light Beryllium atom with the neutron-rich nucleus $^{13}$Be and show that the parameter characterizing the nuclear self-interaction plays an essential role in this case.

## 2. The self-energy function of an unstable nucleus

The natural broadening of spectral line profiles $S(\omega)$ is defined by the probability of a photon being emitted with energy $\omega$ as the atomic system goes from the excited state $|i\rangle$ to a lower state $|f\rangle$:

$$S(\omega) = \frac{dW_i(\omega)}{d\omega} = A'\omega\sum_{\lambda}\int d\Omega_k \left|\langle f;\bar{k},\varepsilon_\lambda |U(t,0)| f\rangle\right|^2 , \tag{1}$$

where $\omega$, $\bar{k}$ and $\varepsilon_\lambda$ are the energy, momentum and polarization of the photon respectively, $A'$ is a normalization factor, $U(t,0)$ is the evolution operator in the representation picture connected with the evolution operator in the Schrödinger picture $U_S(t,0)$ by the following expression: $U(t,0) = e^{iH_0 t} U_S(t,0) e^{-iH_0 t}$. The operator $U_S(t,0)$ is just a Fourier transform of the Green operator:

$$U_S(t,0) = \frac{1}{2\pi}\int_{-\infty}^{\infty} dx \exp(-iz\tau) G(z) . \tag{2}$$

In Ref. [2] it was shown that using the reduction procedure the Green operator can be redifined so that it describes the evolution of particles propagating freely or interacting only with vacuum. This is achieved by means of the insertion of the self-energy function $C(z)$ into the matrix elements of the free Green operator: $\langle m'|\tilde{G}_0(z)|m\rangle = \langle m'|m\rangle (z - E_m - C_m(z))^{-1}$, where $E_m$ is the energy of the system in the state $|m\rangle$ and $C_m(z)$ is the energy of the interaction with vacuum in this state. The condition $z - E_m - C_m(z) = 0$ determines the physical masses of particles. After such reduction the interaction in the system is described by the operator $M(z)$ which involves only the interaction of the particles with each other. The system of equations for the operators $C(z)$ and $M(z)$ was derived in [3] from the generalized dynamical equation (GDE), which in turn is a direct consequence of the first principles of quantum physics. Being equivalent to the Schrödinger equation in the case when the interaction in a quantum system is instantaneous, GDE allows one to extend quantum dynamics to the case of nonlocal-in-time interactions. This equation provides a new insight into many problems in nuclear physics [1,4], atomic physics [5-9] and quantum optics [10,11]. The approach to description of quantum dynamics constructed this way is more general and remains valid even in the case when no Hamiltonian can be constructed as an operator generating dynamics of the system. Moreover, it allows to take into account the instability of a state consistently and to regard it as an energy distribution from the very beginning, without appealing to the eigenvector system of any Hamiltonian.

The Green operator of the atomic system with an unstable nucleus, which undergoes a spontaneous atomic transition with a photon emission, has the following structure:

$$G(z) = \tilde{G}_{DP}(z) M(z) \tilde{G}_f(z) M(z) \tilde{G}_i(z) , \tag{3}$$

where $\langle i|\tilde{G}_i(z)|i\rangle = (z - E_i - C(z - E_i))^{-1}$ is the Green operator of the atomic system in the excited state $|i\rangle$ with the energy $E_i$, $\langle f;\bar{k},\varepsilon_\lambda|\tilde{G}_f(z)|f;\bar{k},\varepsilon_\lambda\rangle = (z - E_f - \omega - C(z - E_f - \omega))^{-1}$ is the Green operator of the atomic system in a lower state $|f\rangle$ with the energy $E_f$ and a photon in the state $|\bar{k},\varepsilon_\lambda\rangle$, $\langle DP|\tilde{G}_{DP}(z)|DP\rangle = (z - E_{DP} - \omega)^{-1}$ is the Green operator of the daughter particles of the nuclear decay with the total energy $E_{DP}$. All the above operators are taken in the Furry picture and the Coulomb interaction between the central nucleus and atomic electrons is already accounted for.

Journal of Physics: Conference Series **613** (2015) 012003 doi:10.1088/1742-6596/613/1/012003

The self energy function of the atom with an unstable nucleus is split up to the self-energy of its electrons $C_{e^-}(z)$ and the self-energy of the nucleus $C_N(z)$: $C(z) = C_N(z) + C_{e^-}(z)$. In general, it has a real and an imaginary part: $C(z) = \Delta E + i\Gamma(z)/2$, where $\Gamma(z)/2 \equiv \operatorname{Im} C(z)$. The real part determines the shift of an energy level, while the imaginary part describes the decay law of the system and in the case of exponential decay coincides with the decay width up to the factor 1/2. Further we will regard the energies of atomic levels with account of the Lamb shift and we will deal with the renormalized masses of the particles. Thus, the real part of the self-energy function is already taken into account and only the imaginary part remains: $C(z) = i\operatorname{Im} C(z)$.

The decay law of the superheavy nucleus is defined by the behaviour of its self-energy function $C_N(z)$ near the point $z = E_N$, $E_N$ being the energy of the supercritical nucleus: $E_N = E_{DP} + \Delta M$, $\Delta M$ being the mass excess. The variation of the self-energy function is generally weak and we can restrict ourselves to the first term of its Taylor expansion: $C_N(z) = C_N(E_N) + dC_N(z)/dz|_{z=E_N}(z - E_N) + \ldots$. Obviously, $2\operatorname{Im} C_N(E_N)$ is just the decay width $\Gamma_0$. The parameter $T_1 = dC_N(z)/dz|_{z=E_N}$ was discussed in [1] and looks as follows: $T_1 = \lambda R$, where $\lambda = \frac{\mu}{\Delta M}$ is the ratio of the reduced mass of the nucleus $\mu$ to its mass excess $\Delta M$. The factor $R$ characterizes the self-interaction of the unstable nucleus. It should be emphasized that this parameter is model independent and provides a fundamental characteristic of the nuclear decay process. Along with the decay width it determines the energy distribution of the unstable state. The value of $R$ determines a nonexponential correction to the decay law. From the solution of the generalized dynamical equation for the nuclear self-energy function [1] it was obtained that $R \sim 1$. With all this, we arrive at the following form of the nuclear self-energy function:

$$2\operatorname{Im} C_N(E_{DP} + \omega) = \Gamma_0 + \lambda \cdot R \cdot \omega. \quad (4)$$

As for the electronic self-function, its dependence on $z$ is not essential in comparison with the nuclear one, so it can be just equated to the decay width $W$ of the corresponding atomic level: $C_{e^-}(z) = iW/2$.

## 3. Spontaneous emission from supercritical atoms

In the case of supercritical atoms $C_{e^-}(z)$ is negligibly small and the total self-energy function is reduced to that of the nucleus: $C(z) \equiv i\operatorname{Im} C_N(z)$. We will investigate the supercritical nucleus on the instance of the double-Uranium giant nuleus. The heavy Uranium nuclei undergo strong Coulomb repulsion and fly apart in a very short time. The daughter particles are two Uranium nuclei. The lifetime of the superheavy nucleus $\tau_{SH}$ plays the key role for the principal probability of the observation of vacuum decay [12] in heavy-ion collision experiments. It is estimated by nuclear physics to be of the order of $\tau_{SH} \sim 10^{-21}$ s [13] (the corresponding decay width is $\Gamma_V \sim 10^5$ eV). At the same time, an assumption was made in [14] that a mechanism can exist which holds the nuclei together for a longer time $\tau_{SH} \sim 10^{-19}$ (the corresponding decay width is smaller: $\Gamma_D \sim 10^3$ eV). The spectroscopic approach described above could help to determine the decay time more precisely.

The spectral line profile for superheavy atoms was derived in [1] and is as follows:

$$S(\omega) = \frac{A}{\Delta M^2 + \Gamma_0^2} \frac{\omega}{(\omega - \Delta M)^2 + (\Gamma_0 + \lambda R \omega)^2 / 4}. \quad (5)$$

The photon emission probability corresponding to $\Gamma_V$ ($\Gamma_D$) will be denoted as $S_V(\omega)$ ($S_D(\omega)$). Here the magnitude of $R$ is not essential in comparison with other parameters ($\Delta M, \Gamma_0$) and its variation will not have a notable effect on the spectrum. Let us explore the influence of the decay width. The numerical values of the parameters in equation (5) are as follows: $\Delta M \approx 9\cdot 10^8 eV$ [14], $\lambda \approx 1.2\cdot 10^2$ for the double-Uranium system [1]. In the optical range ($\omega \sim 1.6-3.2$eV) the photon energy in expression (5) can be neglected in comparison with $\Delta M$ and $\Gamma_0$:

$$P(\omega) = A\omega\left(\Delta M^2 + \Gamma_0^2\right)^{-2}. \tag{6}$$

The photon emission probability is then linearly dependent on the photon energy and the derivative is constant. The numerical value of the derivative differs for the two assumed values of the decay width. This difference manifests itself in the 10th decimal index if $\Gamma_0 \equiv \Gamma_V \sim 10^5$ eV and in the 14th decimal index if $\Gamma_0 \equiv \Gamma_D \sim 10^3$ eV. Measurements of such accuracy are possible owing to the modern optical technologies [15]. In the ultraviolet range the difference manifests itself in the 8th decimal index.

In the soft X-ray band ($\omega \sim 124 \div 5000$ eV) the term $\lambda R\omega$ becomes comparable to $\Gamma_D$ and in the hard X-ray region ($\omega > 5000$ eV) it reaches the value of $\Gamma_V$. The difference between the spectral line profiles manifests itself in the 6th decimal index, the shape of the profile being different as well (figure 1). Thus, the X-ray region is most sensitive from the experimental point of view.

By further increasing the photon energy overcomes the decay width and for gamma photons ($\omega > 124000$) the latter becomes negligible in formula (5). Thus, for high energy photons formula (5) looks as follows:

$$P(\omega) = \frac{A}{\Delta M^2}\frac{\omega}{(\omega - \Delta M)^2} \tag{7}$$

and becomes independent on the decay width. In this part of the spectrum the emission probability is primarily defined by $\Delta M$ and hence can be used for the experimental tests of this parameter.

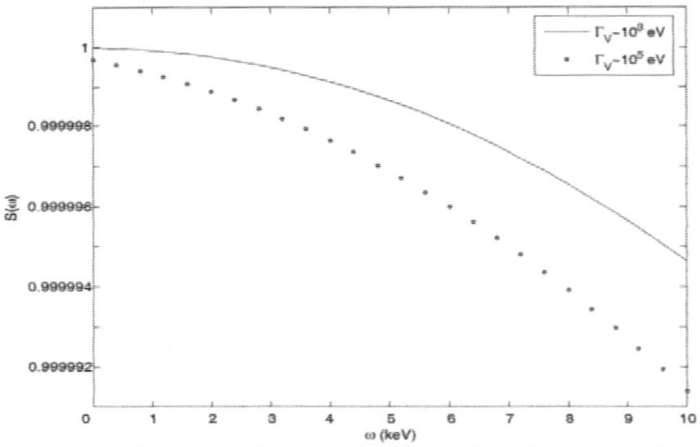


**Figure 1.** The emission probability for the X-ray band.

## 4. Spectrum of the light nucleus $^{13}$Be

Let us consider the electron transition $2p_{1/2} \to 1s_{1/2}$ in a hydrogen-like atom of Beryllium with the neutron-rich nucleus $^{13}$Be. The nucleus $^{13}$Be decays via neutron emission:

$$^{13}_{4}\text{Be} \to {}^{12}_{4}\text{Be} + {}^{1}_{0}\text{n},$$

with the half-life time $\tau_N = 0.51 \cdot 10^{-9}$ s ($\Gamma_0 = 0.13 \cdot 10^{-7}$ eV). The lifetime of the excited atomic state $2p_{1/2}$ is $\tau_A = 0.62 \cdot 10^{-13}$ s (accordingly, $W = 0.11 \cdot 10^{-5}$ eV). And hence the atomic transition occurs before the nucleus decays. But the nuclear instability will influence the atomic spectrum.

The distinction from the previous case is that the electron self-energy is not negligible now and should be taken into account as well, therefore, $C(z) = iW/2 + iC_N(z)$, where $C_N(z)$ is defined by formula (4) as before. The full Green operator of the system is now

$$G(z) = \tilde{G}_f(z) M(z) \tilde{G}_i(z) , \tag{8}$$

where $\langle i|\tilde{G}_i(z)|i\rangle = \left(z - E_i - W/2 - C_N(z - E_i)\right)^{-1}$ is the Green operator of the atomic system in the excited state $|i\rangle$ with the energy $E_i$ and the decay width $W$, $\langle f;\bar{k},\varepsilon_\lambda|\tilde{G}_f(z)|f;\bar{k},\varepsilon_\lambda\rangle = \left(z - E_f - \omega - C_N(z - E_f - \omega)\right)^{-1}$ is the Green operator of the atomic system in a lower state $|f\rangle$ with the energy $E_f$ and a photon in the state $|\bar{k},\varepsilon_\lambda\rangle$. The daughter particles are now a neutron ${}^1_0\mathrm{n}$ and a Beryllium atom with the nucleus ${}^{12}_4\mathrm{Be}$. Their parameters are as follows: $\mu = 0.87 \cdot 10^9$ eV, $\Delta M = 33247823$ eV and hence $\lambda = 26.1$.

Substituting this into (4) and then into (5) we obtain the following form of the spectral line profile:

$$S(\omega) = A \frac{\omega}{\left(\omega - \omega_{fi}\right)^2 + \left(W + \Gamma_0 + \lambda R \omega\right)^2 / 4} , \tag{9}$$

where $\omega_{fi} = 30.2$ eV is the atomic transition energy. Thanks to the fact, that all the intrinsic characteristics for the $^{13}$Be are well known, the only parameter which can be varied here is the factor $R$. The spectral line profiles obtained for different values of this factor are displayed in figure (2).

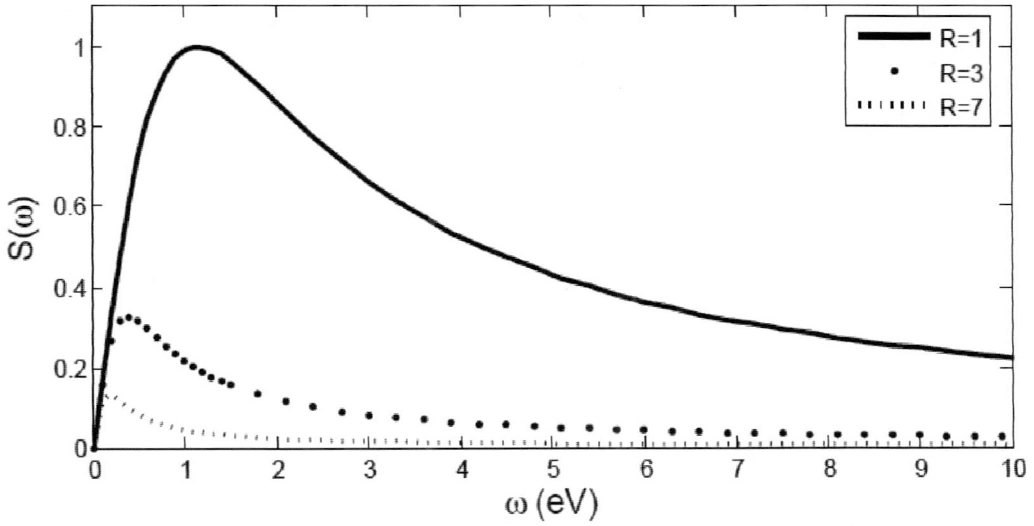


**Figure 2.** The radiation spectrum of the $^{13}$Be atom for the photon energies $\omega \sim 0 \div 10$ eV depicted for different values of the factor $R$.

As it is obvious from the figure, the shape of the spectrum depends dramatically on the value of the factor $R$. This dependence is mostly expressed in the optical and ultraviolet parts of the spectrum, where the highest experimental accuracy is available.

## 5. Summary

The spectra of unstable atoms have been investigated in the framework of generalized quantum dynamics. For very heavy supercritical atoms the spectrum appears to be continuous and to depend mostly on the intrinsic parameters of the nucleus, such as its mass excess and decay width. The X-ray range has been shown to be most sensitive to the magnitude of the decay width of the superheavy nucleus and can be exploited for its experimental determination. The power of the developed approach was demonstrated for light unstable nuclei on the instance of the Beryllium atom with the unstable nucleus $^{13}$Be. It was shown that its spectrum is strongly influenced by the factor depending on the nuclear interaction model and hence the spectrum can be used for the test of theoretical nuclear interaction models.

## 6. References

[1] Gainutdinov R Kh, Mutygoullina A A, Khamadeev M A and Petrova A S 2013 *J. Phys.: Conf. Ser.* **478** 012019
[2] Gainutdinov R Kh 1999 *J. Phys. A: Math. Gen.* **32** 5657
[3] Gainutdinov R K 1989 *J. Phys. A: Math. Gen.* **22** (3) 269
[4] Gainutdinov R Kh and Mutygoullina A A 1997 *Physics of atomic nuclei* **60** 841
[5] Gainutdino R Kh, Khamadeev M A and Salakhov M Kh 2013 *J. Phys.: Conf. Ser.* **478** 012017
[6] Gainutdinov R Kh, Mutygullina A A and Scheid W 2001 *Proc. of SPIE* **4605** 214
[7] Gainutdinov R Kh, Mutygullina A A and Scheid W 2001 *Proc. of SPIE* **4605** 222
[8] Gainutdinov R Kh 1995 *ZETP* **108** 1600
[9] Gainutdinov R Kh 1986 *Optica i Spktroskopiya* **60** 890
[10] Gainutdinov R Kh and Mutygullina A A 2009 *Physica Sripta* T **135** 014021
[11] Gainutdinov R Kh and Mutygullina A A 2013 *J. Phys.: Conf. Ser.* **478** 012018
[12] Heinz S 2005 Topics in Heavy Ion Physics Memorial Symposium for Gerhard Soff (Frankfurt am Main) (Budapest: EP Systema)
[13] Backe H, Senger P, Bonin W, Kankeleit E, Krämer M, Krieg R, Metag V, Trautman N and Wilhelmy J B 1983 *Phys. Rev. Lett.* **50** (23) 1838
[14] Greiner W, Müller B and Rafelski J 1985 *Quantum Electrodynamics of strong fields* (Berlin: Springer)
[15] Hänsch T W 2005 *Nobel lecture* B **10** 1237

Journal of Physics: Conference Series **613** (2015) 012004 doi:10.1088/1742-6596/613/1/012004

# Periodic potential distribution in linear photonic-plasmonic crystals

**A V Koryukin**[1], **A A Akhmadeev**[1] **and M Kh Salakhov**[1,2]

[1] Kazan Federal University, Institute of Physics, 420008, Kremlevskaya St, 18, Kazan, Russian Federation

[2] Tatarstan Academy of Sciences, 420111, Bauman St, 20, Kazan, Russian Federation

E-mail: `Artem.Koryukin@kpfu.ru`

**Abstract.** Photonic-plasmonic crystals are modern materials for nanotechnology and photonics. They have periodic structure and correspondingly periodic potential distribution. We propose a new method of calculation of potential for periodic structures.

## 1. Introduction

Plasmonic nanostructures have found wide application in photovoltaic devices [1]. Photonic-plasmonic crystals (PPCs) [2–8] are used for easily coupled whispering gallery plasmons [9], surface-enhanced Raman spectroscopy [10, 11], coherent fluorescence emission [12], etc. PPC is a hybrid sctructure with combination of dielectric nanophotonic elements with metallic ones supporting surface plasmon polaritons (SPP) [6]. One example of such structures is a self-assembled silica opal covered by gold film [5]. The model of such structure is shown in figure 1. The self-assembled silica opal is a particular case of photonic crystal (PC). PC's are materials with spatial periodicity in their dielectric function (or, equivalently, a periodical index of refraction) [13]. Such periodicity yields a strong modification of the electromagnetic field that manifests itself in new quantum electrodynamical effects [14]. SPP are electromagnetic excitations propagating at the interface between a dielectric and a conductor [15]. Surface plasmons are collective oscillations in electron density at the surface of a metal [16]. SPP properties have various applications in biological and chemical analysis [17]. Calculation of potential is very important for the structures. Since the potential depends on the charge density, the oscillations of surface charge influence the map of the potential. The potential can be obtained experimentally by Kelvin Probe Force Microscopy (KPFM) [18], which allows to obtain an image and to characterize plasmonic modes [19]. It is necessary to calculate the potential for understanding of KPFM imaging. The analytic description of plasmon oscillations in an infinite linear cluster of spherical nanoparticles was developed in [20]. We present a method of calculation of the potential for a structure of arbitrary design, but periodic in one direction. In this connection, we use the Bloch's theorem [21] for periodic medium with the aim of the Coulomb's law simplification. The Kelvin probe method gives the image of potential distribution, which allows to obtain the plasmonic oscillations period. Our method allows to calculate the potential for photonic-plasmonic crystals.

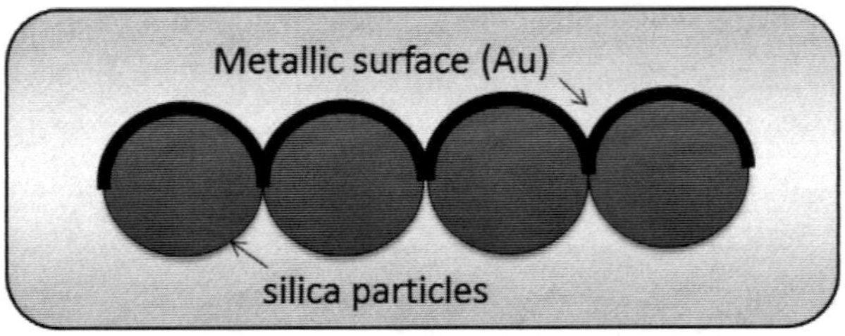


**Figure 1.** Gold film on silica particles as example of photonic-plasmonic crystals.

## 2. Analytical form of potential for periodic structure

The potential can be determined via the Coulomb's law:

$$V(\mathbf{r}) = \int \frac{\rho(\mathbf{r}')}{4\pi\varepsilon_0 |\mathbf{r}-\mathbf{r}'|} \mathrm{d}^3\mathbf{r}', \tag{1}$$

where $V$ is the potential, $\rho$ is the volume charge distribution. Since the physical system has translational symmetry, the Bloch's theorem can be applied. That means that in periodic medium any function $\psi_{\mathbf{k}}(\mathbf{r})$ can be presented in the form:

$$\psi_{\mathbf{k}}(\mathbf{r}) = u_{\mathbf{k}}(\mathbf{r}) e^{i\mathbf{kr}}, \tag{2}$$

where $\psi_{\mathbf{k}}(\mathbf{r})$ is called Bloch wave, $u_{\mathbf{k}}(\mathbf{r})$ is a periodic function with the same periodicity as the medium, $\mathbf{k}$ is the crystal wave vector. Since we research one-dimensional periodic structure in this work, the potential and charge distribution must have Bloch structure along one dimension (axis x):

$$V_k(x,y,z) = \sum_G F_k^{(V)}(y,z;G) e^{i(k+G)x}, \tag{3}$$

$$\rho_k(x,y,z) = \sum_G F_k^{(\rho)}(y,z;G) e^{i(k+G)x}, \tag{4}$$

where $G$ is the vector of reciprocal lattice, $F_k^{(V)}(y,z;G)$ and $F_k^{(\rho)}(y,z;G)$ are the coefficients of expansion for the potential and charge distribution respectively. In accordance with formulas (3) and (4) equation (1) takes the form:

$$\begin{aligned} \sum_G F_k^{(V)}(y,z;G) e^{i(k+G)x} &= \iiint \frac{\sum_{G'} F_k^{(\rho)}(y',z';G') e^{i(k+G')x'}}{4\pi\varepsilon_0 \sqrt{(x-x')^2+(y-y')^2+(z-z')^2}} dx'dy'dz' = \\ &= \frac{1}{4\pi\varepsilon_0} \iint \sum_{G'} F_k^{(\rho)}(y',z';G') e^{i(k+G')x} \int \frac{e^{i(k+G')x''}}{\sqrt{x''^2+a^2}} dx''dy'dz' = \\ &= \frac{1}{4\pi\varepsilon_0} \iint \sum_{G'} F_k^{(\rho)}(y',z';G') e^{i(k+G')x} D_k(y',z';G') dy'dz', \end{aligned} \tag{5}$$

where $D_k(y',z';G)$ is the integral

$$D_k(y',z';G') = \int \frac{e^{i(k+G')x''}}{\sqrt{x''^2+a^2}} dx''. \tag{6}$$

The values $a$, $x'$ are given by the following expressions:

$$a^2 = (y-y')^2 + (z-z')^2, \quad x' = x + x''. \tag{7}$$

Journal of Physics: Conference Series **613** (2015) 012004 doi:10.1088/1742-6596/613/1/012004

Further using the orthogonality properties of exponential functions and Kronecker's delta as follows:

$$\int \sum_G F_k^{(V)}(y,z;G)e^{iGx}e^{-iG''x}dx = \sum_G F_k^{(V)}(y,z;G)\delta_{G,G''} = F_k^{(V)}(y,z;G''), \tag{8}$$

$$\int \sum_{G'} F_k^{(\rho)}(y',z';G')e^{iG'x}e^{-iG''x}dx = \sum_{G'} F_k^{(\rho)}(y',z';G')\delta_{G',G''} = F_k^{(\rho)}(y',z';G''), \tag{9}$$

we obtain:

$$F_k^{(V)}(y,z;G) = \frac{1}{4\pi\varepsilon_0}\iint D_k(y',z';G)F_k^{(\rho)}(y',z';G)dy'dz'. \tag{10}$$

Equation (10) allows to calculate the map of the potential for a structure of arbitrary design with **period** in one direction.

## 3. Calculation for infinitely long, thin periodically charged thread

For example, we can calculate the map of the potential for infinitely long, thin periodically charged thread. Let us write the charge distribution:

$$\rho(x',y',z') = \lambda_0\cos(ikx')\delta(y',0)\delta(z',0). \tag{11}$$

$$\rho(x',y',z') = Re\rho_c(x',y',z'). \tag{12}$$

$$\rho_c(x',y',z') = \lambda_0 e^{ikx'}\delta(y',0)\delta(z',0). \tag{13}$$

We expanded in a series the charge distribution:

$$F_k^{(\rho_c)}(y',z';G') = \lambda_0\delta(y',0)\delta(z',0)\delta_{0,G'}. \tag{14}$$

Substituting (14) in (10), we obtain:

$$F_k^{(V)}(y,z;G) = \frac{1}{4\pi\varepsilon_0}\lambda_0\delta_{0,G}D_k(y',z';G). \tag{15}$$

Taking into account equation (15), the potential can be presented in the form:

$$V_k(x,y,z) = \sum_G F_k^{(V)}(y,z;G)e^{i(k+G)x} = \frac{1}{4\sqrt{2}\pi^2\varepsilon_0}\lambda_0 e^{ikx}D_k(y',z';0), \tag{16}$$

where $D_k(y',z';0)$ is the integral

$$D_k(y',z';0) = \int \frac{e^{ikx''}}{\sqrt{x''^2+a^2}}dx''. \tag{17}$$

The calculation of the real part of $D_k(y',z';0)$ is presented in figure 2. Using this result we can obtain the potential on any distance from thread and with any **k**, excluding k=0.

## 4. Conclusion

We have presented the method of calculation of the potential for a periodic structure with arbitrary design. The method was demonstrated for structures periodic in one direction, but it will be usefull for the two-dimensional case too. It's important because most of photonic-plasmonic crystals have periodicity. The method can be usefull for creating materials with necessary design and parameters.

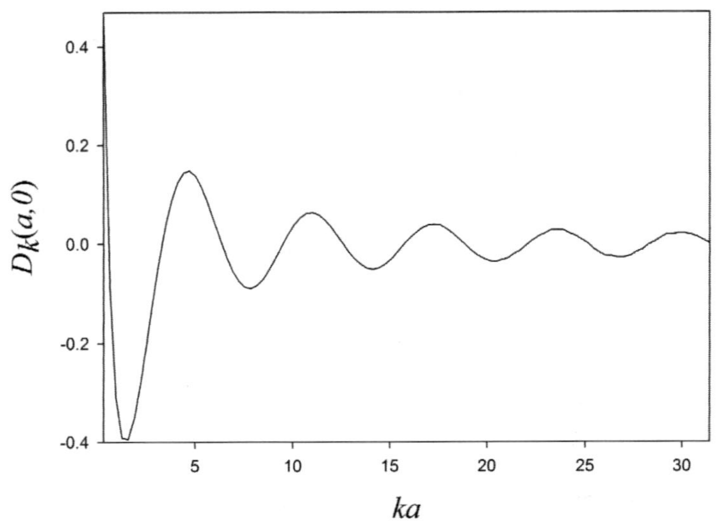


**Figure 2.** Result of calculation of real part of integral in (17) in dimensionless units $k \cdot a$.

## Acknowledgments
This work was funded by the subsidy allocated to Kazan Federal University for the state assignment in the sphere of scientific activities.

## References
[1] Atwater H A and Polman A 2010 *Nature Materials* **9** 205-213
[2] Lopez-Garcia M, Galisteo-Lopez J F, Blanco A,Sanchez-Marcos J, Lopez C and Garcia-Martin A 2010 *Small* **6** (16) 1757-1761
[3] Lopez-Garcia M, Galisteo-Lopez J F, Blanco A, Lopez C and Garcia-Martin A 2010 *Adv. Funct. Mater.* **20** 4338-4343
[4] Galisteo-Lopez J F, Lopez-Garcia M, Lopez C and Garcia-Martin A 2011 *Appl. Phys. Lett.* **99** 083302-1-3
[5] Frederich H, Wen F, Laverdant J, Coolen L, Schwob C and Maitre A 2011 *Optics Express* **19** (24) 24424-24433
[6] Galisteo-Lopez J F, Lopez-Garcia M, Blanco A and Lopez C 2012 *Colloidal Nanoplasmonics* **28** 9174-9179
[7] Frederich H, Wen F, Laverdant J, de Marcillac W D, Schwob C, Coolen L and Maitre A 2014 *Plasmonics* **9** (4) 917-924
[8] Liu T-l, Russel K J, Cui S and Hu E L 2014 *Optics Express* **22** (7) 8219-8225
[9] Cole R M, Sugawara Y, Baumberg J J, Mahajan S, Abdelsalam M and Bartlett P N 2006 *Phys. Rew. Lett.* **97** (13) 137401-1-4
[10] Oh Y-J, Park S-G, Kang M-H, Choi J-H, Nam Y and Jeong K-H 2011 *Small* **7** (2) 184-188
[11] Tsvetkov M Y, Khlebtsov B N, Khanadeev V A, Bagratashvili V N, Timashev P S, Samoylovich M I and Khlebtsov N G 2013 *Nanoscale Research Letters* **8** (250) 1-9
[12] Shi L, Yuan X, Zhang Y, Hakala T, Yin S, Han D, Xhu X, Zhang B, Liu X, Torma P, Lu W and Zi J 2014 *Laser and Photonics Reviews* **8** (5) 717-725
[13] Joannopoulos J D, Johnson S G, Winn J N and Meade R D 2008 *Photonic crystals: molding the flow of light* (Princeton: Princeton University Press) p 286
[14] Gainutdinov R Kh, Khamadeev M A and Salakhov M Kh 2012 *Phys. Rev.* A **85** 053836
[15] Maier S 2007 *Plasmonics: Fundamentals and Applications* (Springer Science+Business Media LLC) p 224
[16] Novotny L and Hecht B 2006 *Principles of nano-optics* (New York: Cambridge Univrsity Press) p 539
[17] Zeng S, Baillargeat D, Ho H-P and Yong K-T 2014 *Chem. Soc. Rev.* **43** 3426-3452
[18] Nonnenmacher M, Oboyle M P and Wickramasingle H K 1991 *Appl. Phys. Lett.* **58** 2921-2923
[19] Cohen M, Shavit R and Zalevsky Z 2014 *Scientific Reports* **4** 4096
[20] Zabkov I V, Klimov V V, Treshin I V and Glazov O A 2011 *Quantum electron* **41** (8) 742-747
[21] Kittel C 1996 *Introduction to Solid State Physics* (Wiley) p 673

Journal of Physics: Conference Series **613** (2015) 012005 doi:10.1088/1742-6596/613/1/012005

# Energy shift experiment in photonic crystal medium

**A A Akhmadeev[1], R Kh Gainutdinov[1], G Hermann[2], M A Khamadeev[1], O V Steryakov[1] and M Kh Salakhov[1,3]**

[1] Institute of Physics, Kazan Federal University, 420008, 18 Kremlevskaya St., Kazan, Russian Federation

[2] Institute of Physics, Justus Liebig University Giessen, Ludwigstrasse 23, Giessen, Germany

[3] Tatarstan Academy of Sciences, 420111, 20 Bauman St., Kazan, Russian Federation

E-mail: `Renat.Gainutdinov@kpfu.ru`

**Abstract.** We propose experimental confirmation of the new effect of the electron mass changes in photonic crystal medium, which was recently predicted. The method consists in measuring the Lamb shift in hydrogen atoms placed in the medium of photonic crystal. We discuss the experimental scheme based on the Lamb and Retherford experiment as well as the requirements for the samples of photonic crystals.

## 1. Introduction

Since the pioneering works of Yablonovitch [1] and John [2] photonic crystals (PCs) attract much attention of researchers. This materials have many potential technological applications ranging from high performance light-emitting diodes and optical fibers to low-power microlasers and optical microchips [3]. These applications are mainly based on the photonic band gap effect. However, band gap is not the only effect that follows from the periodic changing of the refractive index in the photonic crystal. It has been shown that the periodic change of the photon-matter interaction in PC medium gives rise to the fact that the mass of an electron in the PC must differs from its mass in vacuum [4–8]. Actually, this is a novel type QED effect. The origin of the effect under study is the fact that the interaction of a charged particle in the PC with its own radiation field differs from that in vacuum. The matter of fact is that the physical mass of the electron for example is the sum of the "bare" mass and the electromagnetic mass being the electron rest energy arising from processes in which a photon is emitted and then reabsorbed by the electron. The effect should manifest itself in the modification of the spectra of atoms placed in PC medium. For today there are studies where the photoluminescence spectrum of the rare earth ions embedded in inverse opal was observed [9]. The authors detected shifts that was called the Lamb shifts. However, in [7] it was shown that this is not the case and the observed spectral features could be explained by the effect of the change of the mass of an electron. However some straightforward experimental proof of the effect is needed. For this purpose it is important to use the simplest hydrogen atoms, which are unique in their use for comparison between theory and experiment of bound-state energy level structures. We propose to measure the shifts of energy levels of the hydrogen atoms placed in photonic crystal medium. In this paper we discuss ideas for the experimental verification of the effect, based on the Lamb and Retherford experiment [10].

## 2. Influence of environment on the electromagnetic mass of an electron

It was shown in [4] that a new type of quantum-electrodynamical (QED) effect occurs in PCs when the electron rest mass changes. An explanation of this effect is contained in the concept of mass itself: the QED electron rest mass is the sum of its "bare" mass $m_0$ (the mass an electron might have if its interaction with the vacuum were negated) and the electromagnetic mass $m_{em}$, written as

$$m_e = m_0 + m_{em}. \tag{1}$$

Electromagnetic mass $m_{em}$ is a result of an electron always interacting with the vacuum; it creates and then reabsorbs a photon, at which point electron-positron pairs and more complex combinations of virtual particles can arise. Such interaction never ceases.

The problem consists in the fact that the value of the electromagnetic mass of an electron in PC medium should differ from that in vacuum, and therefore the result of adding this electromagnetic mass to its "bare" mass will not be the physical mass. Obviously, the change of the value of the electron mass $\delta m_{pc}$ is the difference between the values of the electromagnetic masses in PC medium $m_{em}^{pc}$ and the electromagnetic mass $m_{em}$ in vacuum

$$\delta m_{pc} = m_{em}^{pc} - m_{em}. \tag{2}$$

Thus, the influence of PC medium on the interaction of an electron with its own radiation field results in the change in its mass: $m_e \to m_{pc} = m_e + \delta m_{pc}$. Here we assume that the electron is in an air void of the PC.

The PC medium correction $\delta m_{pc}$ to the electron rest mass is given by the equation (in the unit system where $\hbar = c = \varepsilon_0 = 1$) [4]

$$\delta m_{pc} = \frac{\alpha}{\mathbf{p}^2\pi^2}\left[\sum_n \int_{FBZ} \frac{d^3k}{\omega_{\mathbf{k}n}^2} \sum_{\mathbf{G}} |\mathbf{p}\cdot\mathbf{E}_{\mathbf{k}n}(\mathbf{G})|^2 - \int \frac{d^3k}{2\mathbf{k}^2} \sum_{\lambda=1}^{2} |\mathbf{p}\cdot\varepsilon_\lambda(\mathbf{k})|^2\right] \tag{3}$$

with $\mathbf{E}_{\mathbf{k}n}(\mathbf{G})$ being the coefficients in the plane-wave expansion $\mathbf{E}_{\mathbf{k}n}(\mathbf{r}) = \sum_{\mathbf{G}} \mathbf{E}_{\mathbf{k}n}(\mathbf{G}) e^{i(\mathbf{k}+\mathbf{G})\cdot\mathbf{r}}$ of the Bloch eigenfunctions $\mathbf{E}_{\mathbf{k}n}(\mathbf{r})$. Here $\mathbf{G}$ is the reciprocal lattice vector of the photonic crystal ($\mathbf{G} = N_1\mathbf{b}_1 + N_2\mathbf{b}_2 + N_3\mathbf{b}_3$ where $\mathbf{b}_i$ are primitive basis vectors of the reciprocal lattice). Thus, in contrast to the Lamb shift, the PC medium correction to the electron mass does not depend on the position of the electron in a PC's air void. At the same time, this correction depends on the direction unit vector $\hat{\mathbf{p}} = \mathbf{p}/|\mathbf{p}|$ of the electron momentum.

The energy of the hydrogen state in free space $|a\rangle = |n, j, l, m\rangle$ may be written as $E_a = m_e + \epsilon_a$ with $\epsilon_a = -\frac{1}{2}\frac{\alpha^2 m_e}{n^2} + O(\alpha^4)$. The frequency $\omega_{ab}$ of the transition between the state $|a\rangle$ and the state $|b\rangle$ is given by $\omega_{ab} = \epsilon_a - \epsilon_b$. Since the mass correction (3) depends on the electron momentum direction the energy gets the correction

$$\langle \delta m_{pc} \rangle = \int d^3p \Psi_{njlm}^*(\mathbf{p}) \delta m_{pc}(\hat{\mathbf{p}}) \Psi_{njlm}(\mathbf{p}), \tag{4}$$

that depends only on the orbital angular momentum and the absolute value of the angular-momentum $z$-component $m$. As a result, the rest energy parts of the total energies of the states $|a\rangle$ and $|b\rangle$ contribute to the transition frequency

$$\omega_{ab}^{pc} = \langle \delta m_{pc} \rangle_a - \langle \delta m_{pc} \rangle_b + \epsilon_a^{pc} - \epsilon_b^{pc}. \tag{5}$$

Thus, we expect that the customary difference between the $2S_{1/2}$ and $2P_{1/2}$ energy levels will change. Moreover each component of the $2P$ doublet will split into two levels corresponding to the values of $m = 0; \pm 1$.

### 3. Experimental evidence of the electron mass changes
The effect under study will very much depend on the structure quality of photonic crystals and many other parameters, including the effective refractive index. Using the model proposed in [4] for the inverse opal PC with parameters $n_{eff} = 1.15$, $\omega_0 = 1$ eV, $h_0 = 0.8$, $\sigma_0 = 0.1$ eV (the last 3 parameters are characterizing the position, the depth and the width of band gap) we expect that the wavelength of the $2S_{1/2} - 2P_{1/2}$ transition will be in the range 200 – 600 $\mu$m. The relation between the observed Lamb shift and the mass correction to the energy levels is shown in the figure 1. Using this relation we can obtain the values of the correction to the mass from experiment.

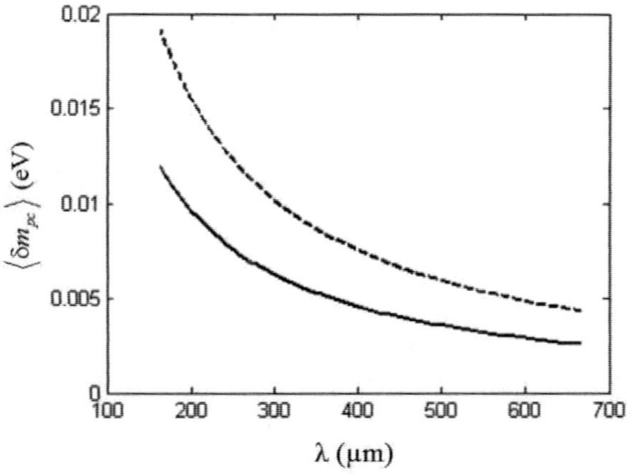


**Figure 1.** The relation between wavelength of the observed $2S_{1/2} - 2P_{1/2}$ Lamb shift and the mass correction to the $S$-state (solid line) and the $P$-state with $m = 0$ (dashed line).

Depending on the PC parameters the shifted lines could be detected in a wide range from a microwaves, up to infrared region (IR). If the energy shifts caused by the effect are in the microwave range, it is natural to use a modified variant of the Lamb and Retherford experiment on the observation of the Lamb shift in hydrogen atoms.

The main distinguishing feature of our experiment is that the hydrogen atoms placed in our voids of the photonic crystal will be exposed to electromagnetic radiation (figure 2). To detect excited atoms it is proposed to use the devices based on electron multipliers, such as the channeltron. Observation of a significant correction to the Lamb shift between the $2S_{1/2}$ and $2P_{1/2}$ energy levels should be an experimental confirmation of the effect under study. The proposed experiment consists of two following steps:

1) Confirmation of the effect of the electron mass changes. To do this, first the Lamb shift of hydrogen atoms will be measured without the photonic crystal. Thus, the frequency of electromagnetic radiation corresponding to the difference between the $2S_{1/2}$ and $2P_{1/2}$ energy levels ($\sim$ 1058 MHz) will be determined and no excited atoms will reach the detector. Then, the photonic crystal is introduced into the scheme in accordance with figure 2. The appearance of excited atoms in the detector will indicate that the value of the Lamb shift is already different from 1058 MHz. As a result, the effect of the electron mass changes in PC medium will be confirmed.

2) Measurement of the new Lamb shift by tuning the frequency of the electromagnetic radiation. As a result of this step the corrections to the energy levels of hydrogen placed in the PC medium will be measured.

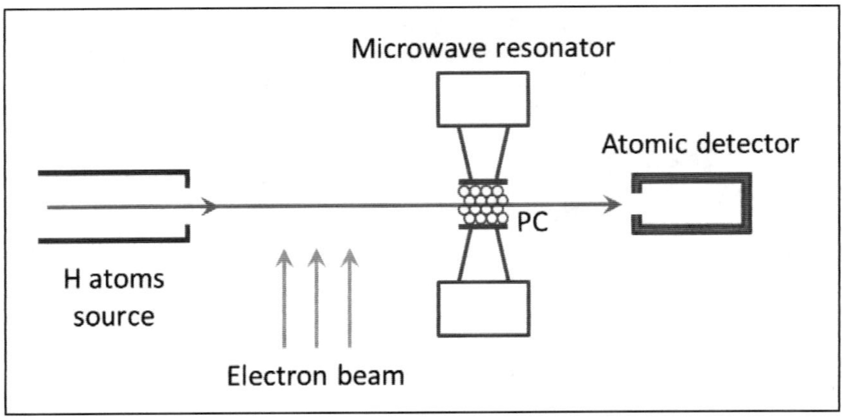


**Figure 2.** Proposed scheme of the experiment.

Certainly there are some issues to be solved. For example, it is necessary to ensure that the atoms from the outside of the photonic crystal will not be exposed to electromagnetic radiation. In this connection it is necessary to concentrate the electromagnetic radiation only in PC using waveguides or antennas. Also, it may be necessary to use IR spectroscopy instead of the microwave generator. In addition, there are many requirements to the photonic crystal, which are necessary for the implementation of our experiment. The structure of the photonic crystal cavities should allow hydrogen atoms to pass through the PC medium. For this purpose, photonic crystals based on inverse opals [11] are best suited. Such materials are obtained by assembling of spherical microparticles in a periodic close-packed structure (synthetic opal) followed by filling of desired substance into the voids between the particles and removing the initial particles. However, it is very difficult to obtain such structures with the required quality.

## 4. Conclusion

The way to experimental confirmation of the effect of the electron mass changes in photonic crystal medium has been proposed. The proposed scheme is based on the Lamb-shift experiment in hydrogen atoms. It is expected that the change in the Lamb shift in photonic crystal will be observed, as well as the corrections to the energy levels in the hydrogen atoms will be measured.

## Acknowledgments

This work was funded by the subsidy allocated to Kazan federal University for the state assignment in the sphere of scientific activities.

## References

[1] Yablonovitch E 1987 *Phys. Rev. Lett.* **58** 2059
[2] John S 1987 *Phys. Rev. Lett.* **58** 2486
[3] Krauss T F and De La Rue R M 1999 *Progr. Quantum Electronics* **23**(2) 51–96
[4] Gainutdinov R Kh, Khamadeev M A and Salakhov M Kh 2012 *Phys. Rev. A* **85** 053836
[5] Gainutdinov R Kh, Salakhov M Kh and Khamadeev M A 2012 *Bull. Russ. Acad. Sci. Phys.* **76**(12) 1301–1305
[6] Gainutdinov R Kh, Khamadeev M A and Salakhov M Kh 2013 *Bull. Russ. Acad. Sci. Phys.* **77**(12) 1440–1443
[7] Gainutdinov R Kh, Khamadeev M A and Salakhov M Kh 2014 *Bull. Russ. Acad. Sci. Phys.* **78**(3) 193–195
[8] Gainutdinov R Kh, Khamadeev M A and Salakhov M Kh 2013 *J. Phys.: Conf. Ser.* **478** 012017
[9] Liu Q et al 2010 *Opt. Lett.* **17** 2898
[10] Lamb W E and Retherford R C 1947 *Phys. Rev.* **72**(3) 241–243
[11] Khokhlov P E, Sinitskii A S and Tretyakov Yu D 2006 *Dokl. Chem.* **408** 61–64

# A Bernasconi model for constructing ground-state spin systems and optimal binary sequences

**A N Leukhin**[1] **and E N Potekhin**[2]

[1]Mari State University, 1, sq. Lenin, Yoshkar-Ola, Mari El Rep, 424000, Russia
E-mail: leukhinan@list.ru,
[2]Volga State University of Technology, 3, sq. Lenin, Yoshkar-Ola, Mari El Rep, 424000, Russia

E-mail: potegor@yandex.ru

**Abstract.** A Bernasconi model for a one-dimensional chain of quantum particles is considered. It is shown that searching for the ground state of such a quantum system is equivalent to searching for optimal binary sequences with minimum energy levels. The second criteria for the optimality of binary sequences with low levels of aperiodic autocorrelation is the minimax criterion, in which the peak level of side lobe (PSL) must be minimal. A review and new results regarding the construction of such binary sequences up to length N=82 are presented.

## 1. Introduction

Classical lattice systems allowing us to describe interatomic potential using finite numbers of values in points defined by the lattice nodes are widely used in statistical physics. The simplest of these is the one-dimensional Ising model [1]. Within such a model, the spin can assume one of two values in each lattice node with number $n$: $s_n = \pm 1$. Let us consider a one-dimensional spin lattice $\vec{S} = (s_0, s_1, \ldots, s_{N-1})$ consisting of $N$ particles with equal distances between neighboring particles.

The energy of such a quantum system consists of pairwise exchange interaction between neighboring atom spins and the interaction of the spins and external magnetic field:

$$E\left(\vec{S}\right) = -\sum_{n,\tau} J_{n\tau} s_n s_\tau - \sum_{n} h_n s_n \,, \tag{1}$$

where indices $n$ and $\tau$ denote the number of lattice nodes, $h_n$ is the strength of the external magnetic field in the $n$th node, and $J_{n\tau}$ is the energy of interaction for spin located in nodes $n$ and $\tau$. With external magnetic field $h = 0$, any energy level is doubly degenerate, since the energy of interaction does not change upon the rotation of all spins. In the simplest case, the energy of interaction can be considered identical for all pairs of neighboring atoms, i.e. $J_{n\tau} = J$. As a result, we obtain the following model for the quantum system's energy:

$$E\left(\vec{S}\right) = -J\sum_{n,\tau} s_n s_\tau \,, \tag{2}$$

In [2], Bernasconi pointed out the connection between the problem of searching for low-energy states of a simplified one-dimensional Ising model (2) and that of constructing binary sequences $\vec{S} = (s_0, s_1, ..., s_{N-1})$ of $N$ length with low levels of autocorrelation.

Aperiodic ACF $\vec{C} = (c_0, c_1, ..., c_{N-1})$ of the binary sequence can be written in the form

$$c_\tau = \sum_{n=0}^{N-1-\tau} s_n \cdot s_{n+\tau} \,,\ \tau = 0,1,...,N-1. \tag{3}$$

In Bernasconi's model, the Hamiltonian of the quantum system of the simplest one-dimensional Ising model is equal to the energy of the sidelobes of the ACF

$$H\left(\vec{S}\right) = \sum_{\tau=1}^{N-1} c_\tau^2 \,. \tag{4}$$

The problem of constructing the simplest quantum system in the form of a one-dimensional Ising model in thus reduced to one of constructing binary sequences with lowest energy of the sidelobes. There are two criteria for the optimality of binary sequences with low levels of aperiodic autocorrelation. The first is the minimax criterion. The peak sidelobe $PSL$

$$PSL\left(\vec{C}\right) = \max_{1 \le \tau \le N-1} |c_\tau| \tag{4}$$

must be minimal

$$MPS = \min_S PSL. \tag{4}$$

The second is the MF (merit factor) coefficient characterizing the ratio of the main frame energy to that of the sidelobes of ACF

$$MF\left(\vec{C}\right) = \frac{N^2}{2\sum_{\tau=1}^{N-1} c_\tau^2} \,. \tag{4}$$

The optimum binary sequences of a given length $N$ has the highest value of the $MF$ coefficient.

Binary sequences that are the optimal according to these criteria are generally different, but for an entire set of lengths they are identical. In this paper we present as known as a new results in the area of constructing optimal MF and MPS sequences. There is known analytical technique to construct sequences with lowest aperiodic autocorrelation, and exhaustive searches have to made in order to find the lowest autocorrelation binary sequence (LABS) for a given length. Many authors have put considerable computational effort to finding binary sequences with lowest or small PSL and the highest or big MF. It should be noted that LABS is problem 005 in the CSPLIB library.

## 2. Related work

Let us mention in short known results to computer search of such binary sequences. There are two search strategies of finding binary sequences of desired length and optimal aperiodic autocorrelations: global and local methods. Local search methods require relatively short time but have shortcomings.

Journal of Physics: Conference Series **613** (2015) 012006 doi:10.1088/1742-6596/613/1/012006

Although they find "reasonable" answers, they can-not guarantee optimality. Also local search methods usually based on initial codes, then determine next codes by using an intermediate criterion. So, if optimum sequences are required, the only solution is global search, for an example, exhaustive search.

The main idea [3] to use an exhaustive search for optimal MF sequences is based on a branch and bound algorithm. Symmetry breaking procedures for identifying equivalent sequences allow the search space to be reduced to approximately one-eighth with the runtime complexity $O\left(1{,}85^N\right)$. Let us introduce the main results of global search for optimal $MF$ binary sequences

- for $2 \le N \le 6$ in 1965 by Lunelly [4];
- for $7 \le N \le 19$ in 1966 by Swinnerton-Dyer, as presented by Littlewood [5];
- for $N \le 32$ by Turyn, as presented by Golay [6] in 1982;
- for $N \le 48$ in1996 by Mertens [3];
- for $N \le 60$ in 2004 (the current record) by Mertens and Bauke [7]

A similar approach has been applied to exhaustive search for optimal $MPS$ sequences. Last modification [8] of such method allows to achieve the runtime complexities $O\left(1{,}42^N\right)$ for $PSL = 2$, $O\left(1{,}57^N\right)$ for $PSL = 3$ and $O\left(1{,}79^N\right)$ for $PSL = 4$. The main results of global search for optimal $MPS$ binary sequences are

- for $N \le 40$ in 1975 by Lidner [9];
- for $N \le 48$ in 1990 by Cohen et al. [10];
- for $N \le 61$ in 1997 by Elders-Boll et all [11];
- for $N = 64$ in 2005 by Coxson and Russo [12];
- for $N \le 68$ in 2012 by Leukhin and Potekhin [13];
- for $N \le 74$ in 2013 by Leukhin and Potekhin [8];
- for $N \le 80$ in 2014 (the current record) by Leukhin and Potekhin [14].

This paper adds to available knowledge for record length of binary $MPS$ sequences and provides numbers of non-equivalent sequences for lengths $N = 81$ and $N = 82$. Also sample binary $MPS$ sequences with highest value of merit factor $MF$ are shown for lengths 2 to 82 and optimal or best known $MF$ sequences for comparing.

## 3. Results of exhaustive search for binary *MPS* sequences

Using our modification of branch and bound algorithm [8] we are able to find all non-equivalent classes of binary $MPS$ sequences for each length $N$ up to $N = 82$. The examples of binary sequences in hexadecimal format for each length $N \in [2;82]$ are presented in table 1:

- the lengths of sequences are presented in the first column shows;
- the values of PSL of optimal $MPS$ sequences are presented in the second column;
- the values of highest merit factor $MF$ of optimal $MPS$ sequences are presented in the third column;
- the optimal $MPS$ sequences with highest merit factor are shown in the forth column;
- the values of PSL of optimal $MF$ sequences (for length $N \in [2;60]$) or best known $MF$ sequences (for length $N \in [2;60]$) are presented in the fifth column;

- the values of highest merit factor *MF* of optimal or best known *MF* sequences are presented in the sixth column;
- the optimal *MF* sequences or best known *MF* sequences are shown in the last column.

**Table 1.** Results of an exhaustive search of binary optimal MPS sequences and binary optimal MF or best known MF sequences.

| Len | PSL | MF | Optimal MPS | PSL | MF | Optimal or best MF |
|---|---|---|---|---|---|---|
| 2 | 1 | 2 | 0 | 1 | 2 | 0 |
| 3 | 1 | 4.5 | 3 | 1 | 4.5 | 3 |
| 4 | 1 | 4 | 2 | 1 | 4 | 2 |
| 5 | 1 | 6.25 | 02 | 1 | 6.25 | 02 |
| 6 | 2 | 2.571 | 0B | 2 | 2.571 | 0B |
| 7 | 1 | 8.167 | 0D | 1 | 8.167 | 0D |
| 8 | 2 | 4 | 16 | 2 | 4 | 16 |
| 9 | 2 | 3.375 | 029 | 2 | 3.375 | 029 |
| 10 | 2 | 3.846 | 076 | 2 | 3.846 | 076 |
| 11 | 1 | 12.1 | 0ED | 1 | 12.1 | 0ED |
| 12 | 2 | 7.2 | 0A6 | 2 | 7.2 | 0A6 |
| 13 | 1 | 14.083 | 00CA | 1 | 14.083 | 00CA |
| 14 | 2 | 5.158 | 019A | 2 | 5.158 | 019A |
| 15 | 2 | 4.891 | 0329 | 3 | 7.5 | 7CCA |
| 16 | 2 | 4.571 | 1DDA | 3 | 5.333 | D2EE |
| 17 | 2 | 4.516 | 0192B | 2 | 4.516 | 0192B |
| 18 | 2 | 6.48 | 0168C | 2 | 6.48 | 0168C |
| 19 | 2 | 4.878 | 07112 | 3 | 6.224 | 7A864 |
| 20 | 2 | 5.263 | 04D4E | 3 | 7.692 | FA2C6 |
| 21 | 2 | 6.485 | 005D39 | 3 | 8.481 | 180CA9 |
| 22 | 3 | 6.205 | 013538 | 3 | 6.205 | 013538 |
| 23 | 3 | 5.628 | 084BA3 | 3 | 5.628 | 084BA3 |
| 24 | 3 | 8 | 31FAB6 | 3 | 8 | 31FAB6 |
| 25 | 2 | 7.102 | 031FAB6 | 3 | 8.681 | 1C7F526 |
| 26 | 3 | 7.511 | 07015B2 | 3 | 7.511 | 07015B2 |
| 27 | 3 | 9.851 | 0F1112D | 3 | 9.851 | 0F1112D |
| 28 | 2 | 7.84 | 1E2225B | 2 | 7.84 | 1E2225B |
| 29 | 3 | 6.782 | 031FD5B2 | 3 | 6.782 | 031FD5B2 |
| 30 | 3 | 7.627 | 03F6D5CE | 3 | 7.627 | 03F6D5CE |
| 31 | 3 | 7.172 | 00E326A5 | 3 | 7.172 | 00E326A5 |
| 32 | 3 | 7.111 | 01E5AACC | 4 | 8 | FEB51CD9 |
| 33 | 3 | 8.508 | 003CB5599 | 3 | 8.508 | 003CB5599 |
| 34 | 3 | 8.892 | 0CC01E5AA | 3 | 8.892 | 0CC01E5AA |
| 35 | 4 | 7.562 | 0CC01E5AA | 4 | 8.39 | 7F64D4AE3 |
| 36 | 3 | 6.894 | 3314A083E | 4 | 7.902 | E073AEDAD |
| 37 | 3 | 6.985 | 006C94A8E7 | 4 | 7.959 | 1FE1E5AA66 |
| 38 | 3 | 8.299 | 003C34AA66 | 3 | 8.299 | 003C34AA66 |
| 39 | 3 | 6.391 | 13350BEF3C | 5 | 7.682 | 7F92D8F1AA |
| 40 | 3 | 7.407 | 2223DC3A5A | 3 | 7.407 | 2223DC3A5A |
| 41 | 3 | 7.504 | 038EA520364 | 5 | 7.782 | 1C3AA66012D |
| 42 | 3 | 8.733 | 04447B874B4 | 3 | 8.733 | 04447B874B4 |
| 43 | 3 | 6.748 | 005B2ACCE1C | 5 | 8.482 | 765B0153862 |
| 44 | 3 | 6.286 | 202E2714B96 | 4 | 7.934 | C1F30422535 |

**Table 1.** Continued

| Len | PSL | MF | Optimal MPS | PSL | MF | Optimal or best MF |
|---|---|---|---|---|---|---|
| 45 | 3 | 6.575 | 02AF0CC6DBF6 | 5 | 8.581 | 1FE4B6270E55 |
| 46 | 3 | 6.491 | 03C0CF7B6556 | 4 | 8.076 | 3FCE1D896CAA |
| 47 | 3 | 7.126 | 069A7E851988 | 4 | 8.181 | 61B0DDD10A50 |
| 48 | 3 | 6.128 | 24AC8847B87C | 4 | 8.229 | C3CC6D1B5402 |
| 49 | 4 | 8.827 | 05E859E984451 | 4 | 8.827 | 05E859E984451 |
| 50 | 4 | 8.17 | 07837FB996B2A | 4 | 8.17 | 07837FB996B2A |
| 51 | 3 | 7.517 | 0E3F88C89524B | 5 | 8.5 | 63C6A1117E4B6 |
| 52 | 4 | 8.145 | 50AE3808C8DB6 | 4 | 8.145 | 50AE3808C8DB6 |
| 53 | 4 | 7.89 | 07C0CFBDB4CD56 | 5 | 8.262 | 19FFC3CCD2D559 |
| 54 | 4 | 7.327 | 116E1DF7D2C6E6 | 5 | 8.331 | 300661C5D66A52 |
| 55 | 4 | 7.451 | 1658A2BC0A133B | 7 | 8.845 | 7FC9392F1B1CAA |
| 56 | 4 | 8.167 | 0C790164F6752A | 4 | 8.167 | 0C790164F6752A |
| 57 | 4 | 7.963 | 01B4DE3455B93BF | 7 | 8.641 | 1C6303AB012B26D |
| 58 | 4 | 8.538 | 008D89574E1349E | 4 | 8.538 | 008D89574E1349E |
| 59 | 4 | 8.328 | 1CAD63EFF126A2E | 5 | 8.49 | 61B4DE3455B93BF |
| 60 | 4 | 8.108 | 119D01522ED3C34 | 6 | 8.257 | CD33A79757B5F80 |
| 61 | 4 | 7.563 | 0024BA568EB83731 | 6 | 8.232 | 1DEFCE255D6C1F92 |
| 62 | 4 | 8.179 | 000C67247C59568B | 4 | 8.179 | 000C67247C59568B |
| 63 | 4 | 9.587 | 1B3412F0501539CE | 5 | 9.587 | 1B3412F0501539CE |
| 64 | 4 | 9.846 | 26C9FD5F5A1D798C | 4 | 9.846 | 26C9FD5F5A1D798C |
| 65 | 4 | 8.252 | 04015762C784EC369 | 5 | 8.802 | 0319EB85AFABF9364 |
| 66 | 4 | 7.751 | 03FEF2CCB0B8CAC54 | 6 | 8.475 | 0F07995533FF996B4 |
| 67 | 4 | 7.766 | 073C2FADC44255264 | 7 | 9.313 | 000F631E169362D55 |
| 68 | 4 | 8.438 | 562B8CA48E0C9027E | 5 | 9.248 | FFD0B564E4D74798E |
| 69 | 4 | 7.988 | 0292582AC6A767CC03 | 6 | 8.688 | 0002F4A9B1B28671 |
| 70 | 4 | 7.313 | 01C2FFD4AF33356596 | 6 | 8.305 | 0F61EC20332A2C5A74 |
| 71 | 4 | 8.105 | 12493BE76A5EE2A3F1 | 5 | 9.165 | 1EC3D840665458B4E9 |
| 72 | 4 | 7.2 | 27C8D6E165A71577FE | 5 | 8.64 | 33CCD84D05FDE1EBD5 |
| 73 | 4 | 8.327 | 012DE781C9167577AB7 | 6 | 8.651 | 01FC677F7A4A2CB972A |
| 74 | 4 | 7.039 | 00ABFA66C560E3094C2 | 6 | 8.029 | 06193EFB2B3DA91F42E |
| 75 | 4 | 7.878 | 0E0038AEB50B59C99B6 | 6 | 8.549 | 00A0AF0E350EE646DB2 |
| 76 | 4 | 7.113 | 2CD864E4AA90B8073DE | 5 | 8.647 | 01415E1C6A1DCC8DB64 |
| 77 | 4 | 6.959 | 066B7BDB752AA6F80E3C | 11 | 8.281 | 00901E934C8CF385AB8A |
| 78 | 4 | 7.548 | 0C4852361E77C0574BAC | 5 | 8.767 | 01C771D006C9293B51AB |
| 79 | 4 | 7.308 | 0028AE35C3A59AC4ED89 | 6 | 9.205 | 7F36491D815A531AA871 |
| 80 | 4 | 6.349 | 01A4F07798EA85AE6C48 | 7 | 9.091 | FFE81E89A8D1C665A9A5 |
| 81 | 4 | 7.594 | 076375955A2F0C69004F9 | 6 | 8.819 | 07A8E3FAA5F7968836676 |
| 82 | 4 | 6.554 | 3CB25D380CE3B7765695F | 5 | 8.918 | 003F70E1112351D4ECB69 |

Size of set of non-equivalent $MPS$ sequences is equal to 2 for the length $N = 81$ and is equal to 1 for the length $N = 82$.

## 4. Conclusion

The problem of constructing the ground states of a quantum system in the form of a sequence of pairwise interacting spins with length $N$ was considered. According to the Bernasconi model the problem can be reduced to one of constructing binary sequences with minimum energy levels of the sidelobes of aperiodic ACF. This paper presents some improvements to previously known results of an

Journal of Physics: Conference Series **613** (2015) 012006 doi:10.1088/1742-6596/613/1/012006

exhaustive search for minimum peak sidelobe sequences. The examples of optimal *MPS* binary sequences up to length $N=82$ with highest merit factor, the examples of optimal *MF* binary sequences up to length $N=60$ and the examples of best known *MF* binary sequences for the lengths from $N \in [61;82]$ are shown.

**Acknowledgments**
This work is supported by the grants: grant of Russian Foundation of Basic Research 12-07-00552 and project of Russian Ministry of Education and Science 1856.

**5. References**

[1] Ising E 1925 *J. Phys.* **31** 253
[2] Bernasconi J 1987 *J. Phys.* **48** 559
[3] Mertens S 1998 *J. Phys. A: Math. Gen* **41** 3731
[4] Lunelli L 1965 *Tabelli di Sequenze (+1,-1) con Autocorrelazione Troncata non Maggiore di 2* (Politecnico di Milano)
[5] Littlewood J E 1968 *J. London Math. Soc.* **41** 367
[6] Golay M J E 1982 *IEEE Trans. Inform. Theory* **IT-28** 543
[7] Mertens S and Bauke H 2004 Ground states of the Bernasconi model with open boundary conditions (Online available <http://odysseus.nat.uni-magdeburg.de/~mertens/bernasconi/open.dat>)
[8] Leukhin A N and Potekhin E N 2013 Optimal peak sidelobe level sequences up to length 74 *Proc. Int. Conf. on European Microwave Conference* (Nuremberg, IEEE Conference publications) p 1807
[9] Lindner J 1975 *Electron. Let.* **11** 507
[10] Cohen M N, Fox M R and Baden J M 1990 Minimum peak sidelobes pulse compression codes *Proceedings of the IEEE International Radar Conference* (Arlington: VA) pp 633-638
[11] Coxson G E and Russo J 2005 *IEEE Trans. Aerosp. Electron. Syst.* **41** 302
[12] Elders-Boll H, Schotten H and Busboom A 1997 Efficient exhaustive search for optimal peak sidelobe binary codes *Proceedings of the IEEE Symposium on Communication and VehicularTechnology* (Benelux) pp 24-31
[13] Leukhin A N and Potekhin E N 2012 Binary sequences with minimum peak sidelobe level up to length 68 (Online available <http://arxive.org/ftp/arxiv/papers/1212/1212.4930>)
[14] Leukhin A N and Potekhin E N 2014 Exhaustive search for optimal minimum peak sidelobe binary sequences up to length 80 *Sequences and Their Applications-SETA 2014 Proceedings of 8th International Conference (Melbourne, VIC, Australia, 24–28 November 2014) (Lecture Notes in Computer Science* vol 8865*)* eds K-U Schmidt and A Winterhof (Springer International Publishing, Switzerland) pp 157–169

Journal of Physics: Conference Series **613** (2015) 012007 doi:10.1088/1742-6596/613/1/012007

# Anti-rotating terms in resonance optics and Lindbladian operators for relaxation

**A M Basharov**

Moscow Institute of Physics and Technology, 141700 Dolgoprudny, Russia
National Research Centre «Kurchatov Institute», 123182 Moscow, Russia

E-mail: basharov@gmail.com; http://basharov.me

**Abstract.** The Lindbladian master equation has been derived for atom and photons of cavity within the theory of open quantum systems based on the effective Hamiltonian and stochastic differential equations. The result obtained differs from the phenomenological one with the emergence of a new additional relaxation operator describing interference processes.

## 1. Introduction

Lindblad [1] has derived a general form of relaxation operator for quantum open systems in Markovian approximation and many authors are using it without any appropriate derivation in their investigations. Such *phenomenological* approaches for studying the relaxation influence on various physical processes in open quantum systems are widespread. Nevertheless, in spite of its general view, the allowance for the Lindbladian relaxation operator obviously implies certain approximations of other terms in the initial Hamiltonian for the problem [2]. Therefore, its allowance with no appropriate approximations is not quite correct, which can lead to erroneous results. This is particularly evident in optical problems where various resonant conditions are satisfied. First of all, to obtain a Lindbladian relaxation operator in optical problems, it is necessary to pass from an initial Hamiltonian to the effective Hamiltonian by means of the unitary transformation of the Hamiltonian of the whole system. Optical problems have a natural small parameter – the ratio of interaction energy to quantum energy and/or energy of quantum transitions. Hence, to obtain a Lindblad operator it is reasonable to use the algebraic perturbation theory based on such unitary transformations and Baker-Hausdorff formula [2]. The principle of term separation of the effective Hamiltonian in resonance optics states that rapidly changing in time terms should be missing in the Dirac picture (interaction representation). This principle was first clearly formulated in works [3, 4], though the similar perturbation theory had been mentioned in the book by Heitler [5], and in the resonance optics some elements of the effective Hamiltonian method were found in work [6]. In work [7], the theory of two-photon and Raman optical resonances is built on the method of the effective Hamiltonian. Work [8] has shown that the principle of absence of rapidly changing in time terms in the effective Hamiltonian provides the natural presentation of any broadband environment of an open system, e.g. a quantized vacuum electromagnetic field in the form of independent quantum noise sources, each of which defines its own Lindbladian relaxation operators of an open system.

Let us demonstrate the statement mentioned above for an open system described in works [9,10]. Let us consider the interaction of a two-level atom with the mode of high-Q cavity. As in [9, 10], the

Journal of Physics: Conference Series **613** (2015) 012007 doi:10.1088/1742-6596/613/1/012007

system loss is supposed to be missing, so both photon and atomic sybsystems have only dephasing described by the Lindblad operators. The master equation for the density matrix $\rho^S$ of an open system in Markovian approximation takes the form

$$\frac{d\rho^S}{dt} = \frac{i}{\hbar}[\rho^S, H^{Eff-S}] - \sum_n \hat{\Gamma}_n \rho^S,\ \hat{\Gamma}_n \rho^S = -\frac{i}{\hbar}[\rho^S, H_n^{Shift-S}] + (\tfrac{1}{2} L_n^{S+} L_n^S \rho^S + \tfrac{1}{2} \rho^S L_n^{S+} L_n^S - L_n^S \rho^S L_n^{S+}).$$

Here, the effective Hamiltonian of an open system $H^{Eff-S}$ is separated from the relaxation operators $\hat{\Gamma}_n$ defined by Lindblad operators $H_n^{Shift-S}$ and $L_n^S$ [2]. However, works [9,10] used the initial Hamiltonian $V_{\text{int}}$ of photon-atom interaction (instead of the effective Hamiltonian $H^{Eff-S}$) and Lindbladian dephasing operators $\hat{\Gamma}_c$ and $\hat{\Gamma}_a$ for photon and atomic subsystems. Also, the works [9,10] discussed the allowance for anti-rotating terms of $V_{\text{int}}$. I would like to emphasize that not only the atom-photon interaction operator $V_{\text{int}}$ has anti-rotating terms, but the interaction operators of subsystems with an environment of an open system also have anti-rotating terms in the initial correct problem formulation. Later in the paper I will prove that the allowance for other anti-rotating terms according to the theory [2] provides additional Lindbladian relaxation operators which have not been taken into account in case of the phenomenological approaches as in works [9,10].

## 2. The simple model for dephasing processes

Let us suppose that there is an atom with two energy levels $|1>$ and $|2>$ with the transition frequency $\omega_0$ between them in the one-mode high-Q cavity with the mode frequency $\omega_c$. To take into account atom and photon dephasing, it is necessary to consider some interaction model of cavity photons and atom with a certain broadband source. The same Lindbladian relaxation operator can be of different origin, but in spite of this the following statement is true - if the Lindblad operator has been accounted correctly, then there occurs a transition from the initial Hamiltonian to the effective Hamiltonian [2]. This leads to the conclusion that one can't take into account anti-rotating terms additionally since they have already been accounted in the effective Hamiltonian. Illustratively, let us take the following model of an open system and its environment, that gives rise to Lindbladian photon dephasing considered in works [9,10].

The objects of an open system will be described by means of the following operators:

1. Cavity photon annihilation operator $c$ obeying the algebra $cc^+ - c^+c = 1$, $[N,c] = -c$, $[N,c^+] = c^+$, where $N = c^+c$ is the operator of the number of photons;
2. Operators $R_\pm$ and $R_3$ describing transitions between atomic levels obey commutation relations of SU(2) algebra: $[R_3, R_\pm] = \pm R_\pm$, $[R_+, R_-] = 2R_3$,, $R_- = |1><2| = R_+^+$, $R_3 = \frac{1}{2}(|2><2| - |1><1|)$.
3. Atom-photon interaction operator $V_{\text{int}} = g(c^+ + c)(R_- + R_+)$ with the coupling constant $g$.

The environment is to be regarded as being composed of non-resonant atoms ("thermostat" atoms) populating the ground energy state, whose parameters have randomly been dispersed in some way. The Hamiltonian of environment atoms is to be designated as $H_b$ and the interaction operator of cavity photons with environment atoms as $V_b$, and

$$H_b = \sum_{b,i} E_i^b \,|\, E_i^b >< E_i^b \,|,\ V_b = (c^+ + c)\sum_{b,ij} d_{ij}^b \,|\, E_i^b >< E_j^b \,|,$$

the upper index $b$ numbers environment atoms and atom parameters.

Thus, the initial Hamiltonian of an open system and environment is as follows:

$$H_{total} = H_a + H_c + H_b + V_{\text{int}} + V_b.$$

According to the general theory of open quantum systems [2], the initial Hamiltonian is subject to unitary transformation

$$\widetilde{H}_{total} = UH_{total}U^+ - i\hbar U \tfrac{d}{dt} U^+ = \sum_{k,m=0.1,2,...} \widetilde{H}^{(k,m)}, \; U = e^{-iS}, \; S = \sum_{k,m=0,1,2,...} S^{(k,m)} - S^{(0,0)},$$

where upper indices indicate the order of term of the transformed Hamiltonian at expansion in terms of coupling constants of cavity mode with a microcavity atom and environment atoms, respectively.

The unique condition for the term selection is the demand for the absence of rapidly changing terms in the interaction representation. If (argument $t$ indicates the interaction representation):

$$\widetilde{H}^{(10)}(t) = \hbar dS^{(10)}(t)/dt + V_{\text{int}}(t), \; \widetilde{H}^{(01)}(t) = \hbar dS^{(10)}(t)/dt + V_b(t), \; \widetilde{H}^{(1,1)}(t) = \hbar dS^{(1,1)}(t)/dt -$$

$$- \tfrac{i}{2}[S^{(0,1)}(t), V_{\text{int}}(t)] - \tfrac{i}{2}[S^{(0,1)}(t), \widetilde{H}^{(1,0)}(t)] - \tfrac{i}{2}[S^{(1,0)}(t), V_b(t)] - \tfrac{i}{2}[S^{(1,0)}(t), \widetilde{H}^{(0,1)}(t)],$$

$$\widetilde{H}^{(2,0)}(t) = \hbar dS^{(20)}(t)/dt - \tfrac{i}{2}[S^{(1,0)}(t), V_{\text{int}}(t)] - \tfrac{i}{2}[S^{(1,0)}(t), \widetilde{H}^{(1,0)}(t)],$$

$$\widetilde{H}^{(0,2)}(t) = \hbar dS^{(0,2)}(t)/dt - \tfrac{i}{2}[S^{(0,1)}(t), V_b(t)] - \tfrac{i}{2}[S^{(0,1)}(t), \widetilde{H}^{(0,1)}(t)],$$

then

$$\widetilde{H}^{(0,0)} = H_a + H_c + H_b, \; \widetilde{H}^{(1,0)} = g(c^+ R_- + cR_+), \; \widetilde{H}^{(0,1)} = 0,$$

and we have the following operators defining the unitary transformation in the interaction representation:

$$S^{(10)}(t) = -ig(cR_- \frac{e^{-i(\omega_c+\omega_0)t}}{\hbar(\omega_0+\omega_c)} - c^+ R_+ \frac{e^{i(\omega_c+\omega_0)t}}{\hbar(\omega_0+\omega_c)}),$$

$$S^{(01)}(t) = ic^+ \sum_{b,ij} | E_i^b >< E_j^b | \frac{e^{i(\omega_c+\omega_{ij}^b)t} d_{ij}^b}{\hbar(\omega_c+\omega_{ij}^b)} - ic \sum_{b,ij} | E_i^b >< E_j^b | \frac{e^{-i(\omega_c-\omega_{ij}^b)t} d_{ij}^b}{\hbar(\omega_c-\omega_{ij}^b)}.$$

It is important to stress that all these terms owe to their appearance to anti-rotating terms in all initial interaction operators. Here, $\omega_{ij}^b = (E_i^b - E_j^b)/\hbar$, for particulars see [3,4].

The effective Hamiltonian of an open system of zero- and first order over coupling constants does not have terms providing phase relaxation of the open-system.

In case of the second order we have terms as in $H_a$ and $H_c$ describing frequency shifts of Lamb type. The former ones are involved into renormalized frequencies $\omega_0$ and $\omega_c$ and are not considered further in the paper. Also, we obtain $\widetilde{H}^{(1,1)} = 0$. There are terms remaining which describe the so-called Stark level shift (alternatively, the Stark interaction [2-4]):

$$\widetilde{H}^{(2,0)} = \eta^{St} c^+ cR_3, \; \widetilde{H}^{(0,2)} = c^+ c \sum_{b,k} \Pi_k^b(\omega_c) | E_k^b >< E_k^b |,$$

$$\eta^{St} = \frac{g^2}{\hbar\omega_c}, \; \Pi_k^b(\omega) = \sum_j \frac{| d_{kj}^b |^2}{\hbar} \left( \frac{1}{\omega_{kj}^b + \omega} + \frac{1}{\omega_{kj}^b - \omega} \right).$$

Parameter $\eta^{St}$ describes the Stark shift for atom levels and makes allowance for $\omega_c \approx \omega_0$. In case of a two-level atom and classical field, this shift is known as the Bloch-Siegert shift [3,4].

### 3. The master equation

Now, to obtain relaxation operators, we need a random process model for $\widetilde{H}^{(0,2)}$. One can substitute the operator $\sum_{b,k} \Pi_k^b(\omega_c) | E_k^b >< E_k^b | dt$ by the increment $dW(t)$ or $dQ(t)$ of classical or quantum random processes [2]. So, using the approach of Ref.[2] for classical and quantum random processes we have derived the following Lindblad operators

$$L_c^S = \sqrt{2\gamma_c} c^+ c\,,\; H_c^{Shift-S} = 0,$$

differing only by the value of relaxation constant $\gamma_c$.

It is to be noted that we considered above the term $\tilde{H}^{(2,0)}$ of the effective Hamiltonian of the second order over the coupling constant of cavity photons and atom. On the one hand, this has proved the allowance for anti-rotating terms in the initial operator of interaction of photons and atom. On the other hand, it is necessary to take into account other "combinational terms". The additional terms making contribution into relaxation determined by the interaction of photons with "thermostat" atoms are as follows

$$\tilde{H}^{(1,2)}(t) = \hbar dS^{(1,2)}(t)/dt - \tfrac{i}{2}[S^{(1,1)}(t),V_b(t)] - \tfrac{i}{2}[S^{(0,2)}(t),V_{\rm int}(t)] - \tfrac{1}{12}[S^{(1,0)}(t),[S^{(0,1)}(t),V_b(t)]] -$$
$$- \tfrac{1}{12}[S^{(0,1)}(t),[S^{(1,0)}(t),V_b(t)]] - \tfrac{1}{12}[S^{(0,1)}(t),[S^{(0,1)}(t),V_{\rm int}(t)]] + \tfrac{1}{12}[S^{(0,1)}(t),[S^{(0,1)}(t),\tilde{H}^{(1,0)}(t)]].$$

This term of the effective Hamiltonian defines an additional relaxation operator $\hat{\Gamma}_{ca}\rho^S$ with Lindblad operators

$$L_{ca}^S = \sqrt{2\gamma_{ca}}(c^+ R_- + cR_+)\,,\; H_{ca}^{Shift-S} = 0.$$

It is important to stress that in [9,10] the unaccounted relaxation parameter $\gamma_{ca}$ is of the same order as accounted parameter $\gamma_c$ in [9,10] in terms of the expansion in powers of coupling constant $g$.

Similarly, consideration can be given to the model of forming phase relaxation of atomic sybsystem of an open system and account can be taken of appropriate «combinational terms» in the effective Hamiltonian of an open system, which will give rise to an additional phase relaxation operator. However, the case of atomic relaxation from the viewpoint of invalid allowance for anti-rotating terms in the phenomenological approach has been considered by the author in other works [11,12], which indicated the impropriety of adiabatic following approximation of the work [13] where the relaxation operator was introduced phenomenologically.

**Acknowledgments**
This work is supported in part by the Russian Foundation for Basic Research (grant No. 13-02-00199-a).

## 4. References

[1] Lindblad G 1976 *Comm. Math. Phys.* **48** 119
[2] Basharov A M 2012 *JETP* **115** 371
[3] Basharov A M 1990 *Photonics. Method of unitary transformation in nonlinear optics* (in Russian, Moscow: Engineering Physics Institute)
[4] Maimistov A I and Basharov A M 1999 *Nonlinear optical waves* (Dordrecht: Kluwer Academic)
[5] Heitler W 1954 *The quantum theory of radiation* (Oxford: Clarendon Press)
[6] Takatsuji M 1970 *Phys. Rev.* B **2** 340
[7] Basharov A M, Maimistov A I and Manykin E A 1983 *Sov. Phys. JETP* **57** 282
[8] Basharov A M 1992 *Sov. Phys. JETP* **75** 611
[9] Werlang T et al 2008 *Phys. Rev.* A **78** 053805
[10] Dodonov A V 2012 *Phys. Scr.* **86** 025405
[11] Basharov A M, Gorbachev V N and Rodichkina A A 2006 *Phys. Rev.* A **74** 042313
[12] Basharov A M and Gorbachev V N 2007 *Optics and Spectroscopy* **102** 585
[13] Klimov A. B. et al 2003 *J. Opt.* B*: Quantum Semiclassical Opt.* **5** 34

Journal of Physics: Conference Series **613** (2015) 012008 doi:10.1088/1742-6596/613/1/012008

# Semiclassical Maxwell-Bloch equations in superradiance theory by input-output formalism of quantum stochastic differential equations

**A M Basharov**

Moscow Institute of Physics and Technology, 141700 Dolgoprudny, Russia
National Research Centre «Kurchatov Institute», 123182 Moscow, Russia

E-mail: basharov@gmail.com; http://basharov.me

**Abstract.** Collective spontaneous relaxation of identical atoms without any quantum correlation has been shown to give a semiclassical description of superradiance. The Maxwell-Bloch equation follows from the N atomic master equation which was derived within the framework of quantum stochastic differential equations.

## 1. Introduction

Collective spontaneous radiation of identical atoms, also known as Dicke superradiance [1], is well described by the Maxwell-Bloch classical equations [2] for a certain range of parameters, the equations derived by means of averaging appropriate quantum equations. In this case, reference is made of the presence of correlations between atoms, due to which semiclassical description is so successful. Meanwhile there are not any quantum correlations [3,4] in an ensemble of two fully excited atoms in the process of collective spontaneous radiation. That is the reason why it is interesting to derive these equations as though they are a consequence of quantum description of spontaneous radiation because in this case the restrictions of semiclassical approach look much clearer. In this work the Maxwell-Bloch semiclassical equations have been derived from quantum stochastic differential equations describing the collective atomic interaction with a vacuum electromagnetic field. An atomic ensemble is localized in a spatial region with dimensions as small as the wavelength. For the given localized atomic system, a vacuum electromagnetic field of a zero photon density is well defined by the Wiener quantum process with the increment algebra [3,4]

$$dB(t)dB^{+}(t)=dt, \quad dB^{+}(t)dB(t)=dB(t)dB(t)=dtdt=dtdB(t)=0\,.$$

## 2. The input-output formalism

The Wiener processes remain unchanged in that sense that values $\widehat{B}(t)=U^{+}(t)B(t)U(t)$ for all meanings $t'\geq t$ are equal to $\widehat{B}(t)=U^{+}(t')B(t)U(t')$. Here $U(t)$ is the evolution operator of a system of localized identical atoms and a vacuum electromagnetic field obeying the quantum stochastic differential equation (QSDU) [3]

$$dU(t)=-\sqrt{\chi}d_{21}\hbar^{-1}dB(t)R_{+}U(t)+\sqrt{\chi}d_{21}^{*}\hbar^{-1}dB^{+}(t)R_{-}U(t)-\tfrac{\chi}{2\hbar^{2}}\,|\,d_{21}\,|^{2}\,R_{+}R_{-}U(t)dt\,,$$

$d_{21}$ is a matrix element of dipole momentum operator for the excited transition between lower - $|1>$ and upper $|2>$ energy levels, $\chi$ is the coupling constant of atom and vacuum electromagnetic field. Other operators are as follows

$$R_3 = \sum_i R_3^{(i)},\ R_\pm = \sum_i R_\pm^{(i)},\ R_3^{(i)} = \frac{1}{2}(|2><2|-|1><1|),\ R_+^{(i)} = |2><1|,\ R_-^{(i)} = |1><2|,$$

$$i = 1,...,N,\ [R_3, R_\pm] = \pm R_\pm,\ [R_+, R_-] = 2R_3.$$

Index $i$ numbers atoms in an ensemble and their characteristics.

Processes of annihilation and creation $B(t)$ and $B^+(t)$ are called *input* processes, whereas processes $\widehat{B}(t)$ and $\widehat{B}^+(t)$ are called *output* processes. The *input* processes describe an electromagnetic field before interacting with an atom system, the output processes characterize the state of an electromagnetic field after interacting with an atom system. The output processes match the criterion of nondestructive measurement: for all $t' \le t$ they commute with any operators $\widehat{A}(t)$ of the atom system: $[\widehat{B}(t'), \widehat{A}(t)] = [\widehat{B}^+(t'), \widehat{A}(t)] = 0$, where $\widehat{A}(t) = U^+(t)AU(t)$, $A$ is an arbitrary operator of atomic system in the Schrödinger representation. It is too difficult to see that output states of an electromagnetic field contain the information of the atomic system. For example,

$$\begin{aligned} d\widehat{B}(t) &= U^+(t+dt)B(t+dt)U(t+dt) - U^+(t)B(t)U(t) = \\ &= (U^+(t)+dU^+(t))(B(t)+dB(t))(U(t)+dU(t)) - U^+(t)B(t)U(t) = dB(t) + \sqrt{\chi}d_{21}^*\hbar^{-1}dtR_-(t), \end{aligned} \tag{1}$$

where it is considered that $[R_+(t), \widehat{B}(t)] = [R_-(t), \widehat{B}(t)] = 0$, since $[R_+, B(t)] = [R_-, B(t)] = 0$.

### 3. One atom and N atomic density matrices

Neglecting correlations (both classical and quantum), one can present an $N$ atomic density matrix $\rho^A(N)$ in the form

$$\rho^A(N) = \rho^{A(1)} \otimes \rho^{A(2)} \otimes ... \otimes \rho^{A(N)}.$$

An N atomic density matrix $\rho^A(N)$ has involved an argument indicating dimensionality of atomic space. In addition,

$$\frac{d}{dt}\rho^A(N) = \frac{d\rho^{A(1)}}{dt}\rho^{A(2)} \otimes ... \otimes \rho^{A(N)} + \rho^{A(1)}\frac{d\rho^{A(2)}}{dt} \otimes ... \otimes \rho^{A(N)} + ...,$$

$$\rho^{A(1)} = Tr_{2...N}(\rho^A(N)),\ Tr(\rho^{A(i)}) = 1,\ Tr_{2...N}(\rho^{A(2)} \otimes ... \otimes \rho^{A(N)}) = 1.$$

Lower indices of tracing indicate the space of atomic states.

In equation (1) the averaging can be carried over the atomic system. Naturally, in such a case, value $Tr_A\{\rho^A\sqrt{\chi}d_{21}^*\hbar^{-1}R_-(t)\}$ is to be regarded as being proportional to the intensity of an electric-field emitted by atoms. This is the classical value. Let us present

$$R_- = R_-^{(1)} + \sum_{i=2}^N R_-^{(i)},$$

and assume that the following approximate equalities are valid in case of a sufficiently large number of atoms

$$Tr_{2...N}(\sum_{i=2}^N R_-^{(i)}\rho^A(N)) \approx \rho^{A(1)}Tr_{2...N}(\sum_{i=2}^N R_-^{(i)}\rho^A(N-1)),$$

$$Tr_{2...N}(\sum_{i=2}^N R_-^{(i)}\rho^A(N-1)) \approx Tr_{1...N}(\sum_{i=1}^N R_-^{(i)}\rho^A(N)) \equiv Tr_A(R_-^{(i)}\rho^A(N)).$$

They have made possible to introduce the classical amplitudes $E^{(-)}$ and $E^{(+)}$ of the intensity of an electric-field emitted by an atomic system as in

$$Tr_{2...N}(\rho^A(N)\sum_{i=2}^{N} R_-^{(i)}) \equiv i\frac{2\hbar}{\chi d_{21}^*}\rho^{A(1)}E^{(-)},\ Sp_{2...N}(\sum_{i=2}^{N} R_+^{(i)}\rho^A(N)) \equiv -i\frac{2\hbar}{\chi d_{21}^*}\rho^{A(1)}E^{(+)}.$$

Then an N-atomic relaxation operator [3-5]

$$\Gamma^{(N)}\rho^A(N) = \tfrac{\chi|d_{21}|^2}{2\hbar^2}(R_+R_-\rho^A(N)+\rho^A(N)R_+R_- - 2R_-\rho^A(N)R_+),$$

can be transformed with account of the following equations

$$Tr_{2...N}(\rho^A(N)\sum_{i=2}^{N} R_+^{(i)}\sum_{i=2}^{N} R_-^{(i)} + \sum_{i=2}^{N} R_+^{(i)}\sum_{i=2}^{N} R_-^{(i)}\rho^A(N) - 2\sum_{i=2}^{N} R_-^{(i)}\rho^A(N)\sum_{i=2}^{N} R_+^{(i)}) = 0,$$

The expression

$$\frac{\chi|d_{21}|^2}{2\hbar^2}Sp_{2...N}(\rho^A(N)R_+^{(1)}R_-^{(1)} + R_+^{(1)}R_-^{(1)}\rho^A(N) - 2R_-^{(1)}\rho^A(N)R_+^{(1)}) =$$

$$= \frac{\chi|d_{21}|^2}{2\hbar^2}(\rho^{A(1)}R_+^{(1)}R_-^{(1)} + R_+^{(1)}R_-^{(1)}\rho^{A(1)} - 2R_-^{(1)}\rho^{A(1)}R_+^{(1)}) = \Gamma^{(1)}\rho^{A(1)}$$

is an ordinary one atom relaxation operator [3-5]. As a result, we have obtained the ordinary Bloch equations

$$(\frac{d\rho^{A(1)}}{dt} + \Gamma^{(1)}\rho^{A(1)})_{21} = id_{10}E^{(-)}\hbar^{-1}(-\rho_{22}^{A(1)} + \rho_{11}^{A(1)}),$$

$$(\frac{d\rho^{A(1)}}{dt} + \Gamma^{(1)}\rho^{A(1)})_{12} = -id_{21}^*E^{(+)}\hbar^{-1}(\rho_{11}^{A(1)} - \rho_{22}^{A(1)}),$$

$$(\frac{d\rho^{A(1)}}{dt} + \Gamma^{(1)}\rho^{A(1)})_{22} = i(-d_{10}^*E^{(+)}\rho_{21}^{A(1)} + \rho_{12}^{A(1)}d_{10}E^{(-)})\hbar^{-1},$$

$$(\frac{d\rho^{A(1)}}{dt} + \Gamma\rho^{A(1)})_{11} = i(-\rho_{12}^{A(1)}d_{21}E^{(-)} + d_{21}^*E^{(+)}\rho_{21}^{A(1)})\hbar^{-1}.$$

The advancement of describing superradiance in a semiclassical manner is likely to specifically rest upon the absence of quantum correlations [3,4], though it is the formation of correlations in an atomic system that is frequently discussed in the process of explaining superradiance. The analogous analysis of collective relaxation operator is unknown to the author for the case of squeezed electromagnetic thermostat. Thus, the main condition of validity of using Maxwell-Bloch classical equations to describe superradiance is the presence of a sufficiently large number of atoms.

**Acknowledgments**
This work is supported in part by the Russian Foundation for Basic Research (grant No. 13-02-00199-a).

## 4. References

[1] Dicke R 1954 *Phys.Rev.* **93** 99
[2] Benedict M G, Ermolaev A M, Malyshev V A, Sokolov I V and Trifonov E D 1996 *Super-radiance: Multiatomic Coherent Emission* (Bristol and Philadelphia: IOP)
[3] Basharov A M 2002 *JETP Lett.* **75** 123
[4] Basharov A M 2002 *JETP* **84** 1070
[5] Gardiner C W and Zoller P 2000 *Quantum noise* (Berlin, Heidelberg: Springer-Verlag)

Journal of Physics: Conference Series **613** (2015) 012009 doi:10.1088/1742-6596/613/1/012009

# Optimization of a heralded single-photon source with spatial and temporal multiplexing

**I Z Latypov[1], A V Shkalikov[1] and A A Kalachev[1,2,3]**
[1]Zavoisky Physical-Technical Institute, 10/7, Sibirsky tract str., 420029, Kazan, Russia
[2]Kazan Federal University, 18, Kremlevskaya str., 420008, Kazan, Russia
[3]Institute for Informatics of Tatarstan Academy of Sciences, 36a, Levobulachnaya, 420011, Kazan, Russia

E-mail: a.a.kalachev@mail.ru

**Abstract.** The properties of a heralded single-photon source with temporal and spatial multiplexing are studied with the aim to maximize its efficiency for a given value of the second-order zero-time autocorrelation function. We show that the variable time delay, which is used for temporal multiplexing, can be optimized so that the mean number of photon passes through the switches and the total number of switches are respectively reduced to ~ $\log_2 N$ and ~ $(1/2)\log_2 N$, where $N$ is the temporal multiplexing degree. The total efficiency of such an optimized source is calculated for typical switching losses and the autocorrelation function is calculated in the presence of the detector dark-count noise.

## 1. Introduction

Generating single-photon states via spontaneous parametric down-conversion (SPDC) or spontaneous four-wave mixing (SFWM) is currently the most convenient experimental technique, which has been widely used in quantum optics and quantum information [1]. The photon number correlation between the down-conversion fields, which are usually referred to as signal and idler fields, allows one to herald the existence of one photon by detection of its partner. Such conditional preparation of single-photon states has several advantages. First, it allows one to generate a single photon into a given spatial-temporal mode of the electromagnetic field by using an appropriate filter during the detection process. Second, the nonlinear optical phenomena make it possible to generate pure single-photon states in a wide spectral range at room temperatures, provided that the nonlinear medium has specific dispersion properties. Third, the contribution of the vacuum state into the output field is removed by the heralding procedure. On the other hand, there are two significant drawbacks: nondeterministic generation of the heralded photons and nonzero contribution of multiphoton states into the output field. To approach deterministic source and simultaneously reduce the multiphoton state contribution, one can take advantage of spatial [2–6] or/and temporal [7–11] multiplexing. In both cases, several processes of photon pair generation are combined in a single source so that heralding probability can be increased without increasing generation rate in each process thereby achieving high fidelity of the source. Both methods are experimentally demonstrated [4–6, 9]. Recently a combination of spatial and temporal multiplexing has also been proposed [12], and different multiplexing schemes have been studied numerically [13–15].

In the present work, we show new possibilities for optimization of temporal multiplexing in a heralded single-photon source and analyze different combinations of temporal and spatial multiplexing under typical experimental conditions. In addition, we calculate the autocorrelation function of the output field in the presence of the detector dark-count noise.

## 2. Multiplexing

The spatial multiplexing means generation of photon pairs in a series of nonlinear sources so that heralding photon in each pair can be detected in each source and only one heralded photon can be routed to the output (figure 1). The spatial multiplexing degree $M$ is equal to the number of nonlinear sources. From the view point of losses, the optimal scheme of the router is a system of switching couplers $2\times1$, the number of which increases with the multiplexing degree as $\log_2 M$ [3].

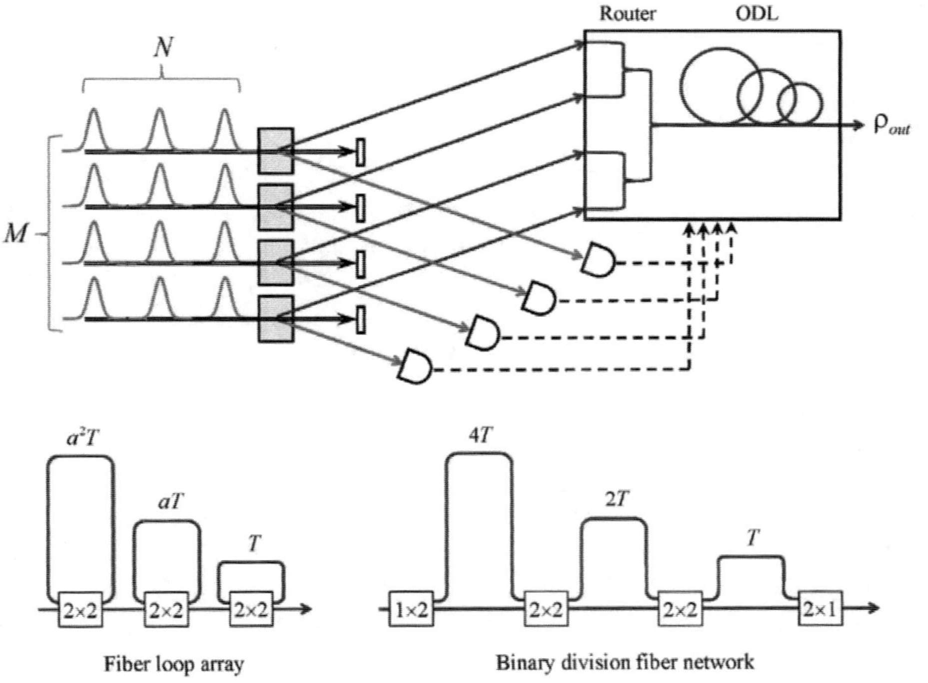


**Figure 1.** Illustrating heralded single-photon source with spatial and temporal multiplexing (above) and possible schemes of a variable optical delay line (below).

The temporal multiplexing involves repeated application of a pump pulse and a variable optical delay line (ODL) for the heralded photons. The temporal multiplexing degree $N$ is determined by the number of pump pulses that are used for generating a single photon. These pulses form a multiplexing cycle. Regarding the variable ODL, a fiber loop array (FLA) can be used. A free-space variant of such a variable time delay has recently been demonstrated [16]. Another convenient architecture is a binary division fiber network (BDFN) [10, 11], which is used in commercial devices. In the present paper, we consider the first variant and show that the system of storage loops can be optimized so that its efficiency becomes higher, while the total number of switches becomes smaller, than that of binary division network. This is possible because the ratio between the fiber loop lengths $a$ may be arbitrary and not necessarily equal to 2 or 10. The optimal value of the parameter $a$ depends on the temporal multiplexing degree.

### 3. Basic properties of the heralded source with multiplexing

Let us consider a heralded single-photon source where both spatial and temporal multiplexing are used. In what follows, we estimate the maximum efficiency of such a source, which can be achieved under typical experimental conditions.

The density operator of the output field may be written as

$$\rho_{out} = \left(1 - \sum_n P(n)\right)|0\rangle\langle 0| + \sum_n P(n)|n\rangle\langle n|, \tag{1}$$

where $P(n)$ is the probability of observing $n$-photon state at the output by the end of a multiplexing cycle ($n > 0$). If the signal and idler photons are emitted into a single spatial-temporal mode, the state vector of the biphoton field takes the form:

$$|\psi\rangle = c_0|0_s 0_i\rangle + c_1|1_s 1_i\rangle + \ldots, \quad c_n = \left[\text{th}(r)\right]^n / \text{ch}(r), \tag{2}$$

where the indices $s$, $i$ denote the modes of the signal and idler field, respectively, and $r$ is the pump parameter. Let $p_n$ be the detection probability provided that the detector interacts with the $n$-photon state. In the case of a bucket detector, which does not discriminate between photon numbers, $p_n = 1-(1-\eta)^n$, where $\eta$ is the detector efficiency. It is easy to show that the probability of heralding after a single pump pulse is equal to

$$P^M_{herald} = 1-(1-P_{herald})^M, \quad P_{herald} = p_1|c_1|^2 + p_2|c_2|^2 + \ldots \tag{3}$$

Here $M$ is the spatial multiplexing degree. Let the cycle of temporal multiplexing consist of $N$ pump pulses separated in time by $T$. If a correlated photon pair is emitted during $i$th pump pulse, the signal photon delay should be equal to $(N-i)T$. As a result, each signal photon appearing during the cycle is emitted as it would be generated by the last pump pulse. Let us assume that the output photon in each cycle corresponds to the last heralding pulse, which leads to the minimum losses. Then

$$P(n) = \left(P^M_{herald}/P_{herald}\right)\sum_{i=1}^{N}\left(1-P^M_{herald}\right)^{N-i}\sum_{k=n}^{\infty} t^n(1-t)^{k-n}\binom{k}{n} p_k |c_k|^2, \tag{4}$$

where

$$t = (\eta_L)^{N-i}(\eta_{SW})^{N_{SW}(i,N)+\log_2 M}. \tag{5}$$

Here $\eta_L$ is the efficiency of a shortest fiber loop providing the delay time $T$, and $\eta_{SW}$ is the efficiency of the switching elements that are used in the FLA and in the router. The number of signal photon passes through the switching elements in the FLA for a given delay time $(N-i)T$ is denoted as $N_{SW}(i,N)$. For example, if the FLA consists of three loops delaying on time $a^0T$, $a^1T$ and $a^2T$ so that $N-i = A_0a^0 + A_1a^1 + A_2a^2$, then $N_{SW}(i,N) = A_0 + A_1 + A_2 + 3$. The number of passes through the switching elements in the router is assumed to be equal to $\log_2 M$. The summation over $k$ in equation (4) takes into account the photon number redistribution due to the losses in the FSSL and in the router [8].

Now we are able to calculate the basic parameters of the single-photon source. In particular, the total efficiency, i.e., the probability of generating a single-photon state in a single multiplexing cycle, is equal to $E = P(1)$. Another important parameter is the heralding efficiency $E_{herald} = P(1)/P^{MN}_{herald}$, which illustrates the efficiency of the optical path from the nonlinear media to the output. Here $MN$ is the total multiplexing degree. The theoretical value of the second-order zero-time autocorrelation function $g^{(2)}(0)$ is calculated as

$$g^{(2)}_{theor}(0) = \left(\sum P'(n)n\right)^{-2}\sum P'(n)n(n-1), \tag{5}$$

where $P'(n)=P(n)/\sum P(n)$. The corresponding experimental value, which is measured using a 50/50 beam splitter and two detectors *A* and *B*, is calculated as

$$g^{(2)}_{\exp}(0)=\left(\sum P'(n)P^n_A\right)^{-2}\sum P'(n)P^n_{AB}, \tag{6}$$

where

$$P^n_{AB}=1+(1-\eta_A)^n-2(1-\eta_A/2)^n-4p_{dc}\left[(1-\eta_A)^n-(1-\eta_A/2)^n\right] \tag{7}$$

is the coincidence count probability with the *n*-photon state at the input of the beam splitter, and

$$P^n_A=1-(1-p_{dc})(1-\eta_A/2)^n \tag{8}$$

is the count probability of a single detector. Here $p_{dc}$ is the dark count probability, which is assumed to be equal for the detectors. So is the efficiency $\eta_A$.

## 4. Basic results

Numerical simulations on the basis of equation (4) show that the maximum efficiency of the single-photon source is achieved at the optimal value of the parameter *a*, which depends on the multiplexing degree and not necessarily equal to 2. To minimize the number of photon passes through the switches we can take $a=\sqrt[s]{N+1}$, where *s* is the number of storage loops. The optimal number of sources for given losses is found numerically by finding the multiplexing degree which maximizes the total efficiency for a given value of $g^{(2)}(0)$. It should be noted that such optimization is different from the usual approach which does not take into account the value of the autocorrelation function. In order to compare different combinations of temporal and spatial multiplexing, we consider only multiplexing degrees which are powers of 2. To be more specific, we take $\eta_{SW}=0.891$ (0.5 dB insertion loss), $\eta_L=0.999$ ($T=100$ ns in a fiber loop with 0.2 dB/km loss), and $g^{(2)}(0)=0.05$. In this case, the maximum efficiency is achieved when $MN\simeq 256$ with the optimum value of *s* depending on the temporal multiplexing degree *N* (table 1).

**Table 1.** Total efficiency *E* for different combinations of temporal and spatial multiplexing. The result of numerical simulation on the basis of equation (4) with the following values of parameters: $\eta=0.5$, $\eta_L=0.999$, $\eta_{SW}=0.891$.

| *M* | *N* | *E* | | | |
|---|---|---|---|---|---|
| | | $s=1$ | $s=2$ | $s=3$ | $s=4$ |
| 1 | 256 | 0.12 | 0.31 | 0.38 | ***0.38*** |
| 2 | 128 | 0.20 | 0.36 | ***0.40*** | 0.39 |
| 4 | 64 | 0.30 | 0.40 | ***0.41*** | 0.37 |
| 8 | 32 | 0.38 | ***0.42*** | 0.41 | 0.38 |
| 16 | 16 | ***0.43*** | 0.42 | 0.40 | 0.36 |

The dependence can be approximated as $s_{opt}=\lfloor(1/2)\log_2 N\rfloor$. Under such conditions, the mean number of photon passes through the switches proves to be about $\log_2 N+1$. On the other hand, when using the BDFN, we obtain $E=\eta_{SW}^{\log_2 M+\log_2 N+1}=0.35$, which never exceeds the efficiency of the optimized FLA. In addition, the total number of switches in the optimized FLA is two times smaller than $\log_2 N+1$ which is needed for the BDFN. Figure 2 illustrates the basic parameters as functions of the mean photon number $\langle n\rangle\approx r^2\approx P(1)$ under expected experimental conditions.

Regarding the autocorrelation function, the value $p_{dc} = 10^{-5}$ corresponds to the dark count rate 1 kHz and detection window 10 ns. It can be seen from figure 2 that such a dark count rate reveals itself only for high fidelity, when $g^{(2)}(0) < 0.01$. In this case, a measured value of the autocorrelation function may differ significantly from an expected theoretical one. However, increasing $p_{dc}$ up to $10^{-4}$ gives an error 50% when the second-order zero-time autocorrelation function is equal to 0.01.

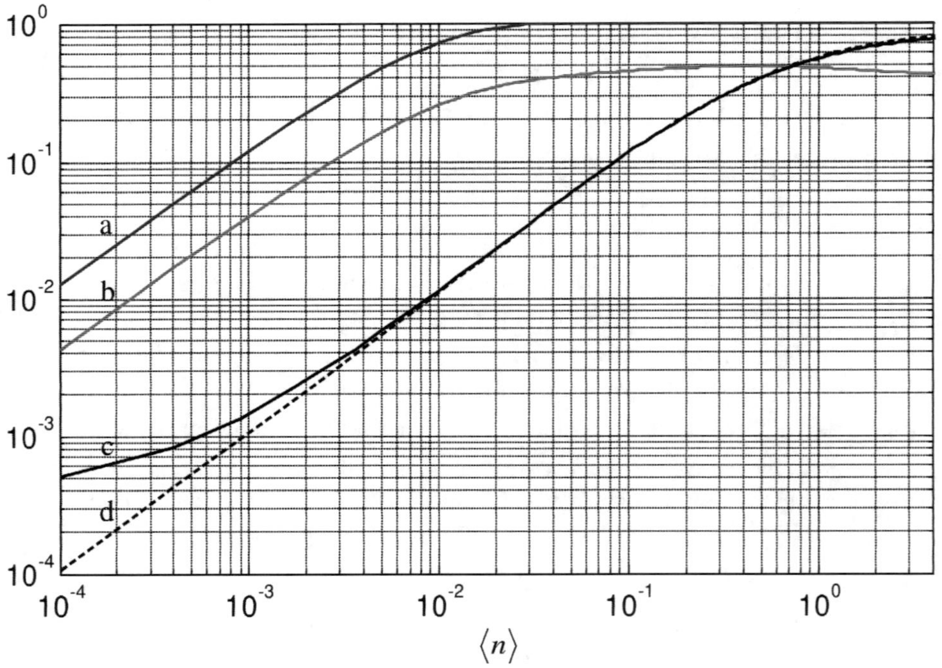


**Figure 2.** Heralding efficiency (a), total efficiency (b), experimental $g^{(2)}(0)$ (c) and theoretical $g^{(2)}(0)$ (d) as functions of the mean photon number $\langle n \rangle$. The result of numerical simulation on the basis of equation (4) with the following values of parameters: $N = 64$, $M = 4$, $\eta = 0.5$, $\eta_A = 0.2$, $\eta_L = 0.999$, $\eta_{SW} = 0.891$, $a = 4$ ($s = 3$), $p_{dc} = 10^{-5}$.

## 5. Conclusion

We have studied basic properties of a heralded single-photon source with temporal and spatial multiplexing. We found that the efficiency of a variable optical delay line, which is used for temporal multiplexing, can be optimized so that the total number of switches is reduced twice compared with the binary division network, while the efficiency becomes higher. In this respect, the optimized optical delay line seems to be the easiest one for implementation. According to the numerical simulations, a single-photon source with the efficiency about 40% and the second-order zero-time autocorrelation function 0.05 may be realized under typical experimental parameters.

## Acknowledgments

This research was supported by the Russian Foundation for Basic Research (Grant No. 14-12-0719214-ofi).

## References

[1] Eisaman M D, Fan J, Migdall A and Polyakov S V 2011 *Rev. Sci. Instrum.* **82** 071101
[2] Migdall A L, Branning D and Castelletto S 2002 *Phys. Rev.* A **66** 053805
[3] Shapiro J H and Wong F N 2007 *Optics Letters* **32** 2698

[4] Ma X S, Zotter S, Kofler J, Jennewein T and Zeilinger A 2011 *Phys. Rev.* A **83** 043814
[5] Collins M J, Xiong C, Rey I H, Vo T D, He J, Shahnia S, Reardon C, Krauss T F, Steel M J, Clark A S and Eggleton B J 2013 *Nature Communications* **4** 2582
[6] Meany T, Ngah L A, Collins M J, Clark A S, Williams R J, Eggleton B J, Steel M J, Withford M J, Alibart O and Tanzilli S 2014 *Laser and Photonics Reviews* **8** L42
[7] Pittman T B, Jacobs B C and Franson J D 2002 *Phys. Rev.* **66** 042303
[8] Jeffrey E, Peters N A and Kwiat P G 2004 *New Journal of Physics* **6** 100
[9] Peters N A, Arnold K J, VanDevender A P, Jeffrey E R, Rangarajan R, Hosten O, Barreiro J T, Altepeter J B and Kwiat P G 2006 *Proc. SPIE* **6305** 630507
[10] Mower J and Englund D 2011 *Phys. Rev.* A **84** 052326
[11] Schmiegelow C T and Larotonda M A 2013 *Appl. Phys.* B **116** 447
[12] Glebov B L, Fan J and Migdall A 2013 *Appl. Phys. Lett.* **103** 031115
[13] Mazzarella L, Ticozzi F, Sergienko A V, Vallone G and Villoresi P 2013 *Phys. Rev.* A **88** 023848
[14] Adam P, Mechler M, Santa I and Koniorczyk M 2014 *Phys. Rev.* A **90** 053834
[15] Bonneau D, Mendoza G J, O'Brien J L and Thompson M G 2014 Effect of loss on multiplexed single-photon sources *Preprint* quant-ph/1409.5341
[16] Christensen B G, McCusker K, Goggin M, Crimmins K and Kwiat P 2011 Free-space photon storage with variable time delays *CLEO:2011 – Laser Applications to Photonic Applications, OSA Technical Digest (CD)* (Optical Society of America, 2011) paper QThJ2

Journal of Physics: Conference Series **613** (2015) 012010 doi:10.1088/1742-6596/613/1/012010

# Strong focusing higher-order laser modes: transverse and longitudinal optical fields

**A V Kharitonov and S S Kharintsev**

Department of Optics and Nanophotonics, Institute of Physics, Kazan Federal University, Kremlevskaya, 16, Kazan, 420008, Russian Federation

E-mail: antonharitonov91@gmail.com

**Abstract**. The distribution of transverse and longitudinal optical fields in tightly focused higher-order laser beams is investigated. Polarization-dependent fingerprints of transverse and longitudinal optical fields are experimentally captured by means of photoinduced surface deformations in azobenzene polymer thin films.

Most important problems in modern photonics are fabrication, visualization and characterization of photonic materials and metamaterials [1,2,3]. Single molecule sensitive optical tools such as tip-enhanced Raman scattering, nano infrared absorption, time-resolved coherent anti-Stocks Raman scattering and others play a crucial role [4,5,6,7]. A number of optical techniques uses tightly focused (N.A.>1) higher-order laser modes to access longitudinal optical fields, which are directed towards the wave vector. The paraxial approximation, which states that electric and magnetic fields have to be transverse to the propagation direction, becomes insufficient. Different approaches have been developed for a theoretical description of strongly focused light [8,9,10].

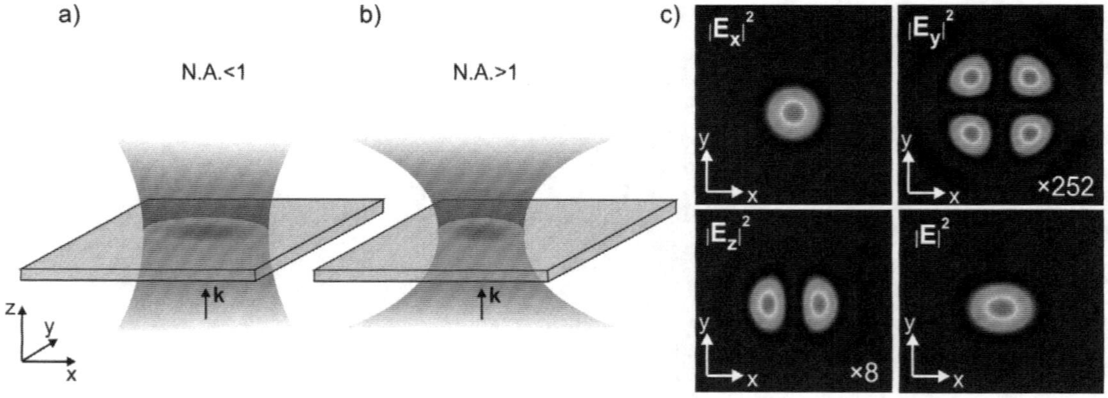


**Figure 1.** A weakly focused (a) and highly focused (b) laser light. (c) The intensity of different electric field components in the focal plane (N.A.=1.4); image sizes $2\lambda \times 2\lambda$. Before focusing, the incident light is polarized towards the x-axis.

Conventional optical spectroscopy and microscopy utilize a weakly focused linearly polarized laser light (figure 1 (a)). In order to increase the spatial resolution, a highly focused light (figure 1 (b)) is used. Figure 1 (c) shows the intensity of different electric field components in the focal plane. As shown in the figure, the field in the focal plane is not purely linearly polarized. In particular, two lobes of the longitudinal field, polarized along the *z* axis, arise on the rims of the laser spot. The maximum intensity of the longitudinal electric field is 8 times smaller as compared to the transverse *x*-component. The development of technology, which allows one to control the energy distribution between field components and their spatial position, is important in optical applications. For example, in single molecule microscopy, which uses a linearly polarized light, the exciting field $\mathbf{E}$ is mainly polarized in the sample plane. The fluorescence intensity is proportional to $|\mathbf{E}\cdot\boldsymbol{\mu}|^2$, where $\boldsymbol{\mu}$ is a dipole moment of the molecule. Thus, it is possible to detect only those molecules whose dipole moments are oriented in the plane of a substrate. Optical schemes in which longitudinal optical fields are used, reasonably enrich single molecule spectroscopy [11,12,13]. These measurements are important in biology [14], biochemistry [15] and quantum optics [16]. Longitudinal field modes have found applications in high resolution near-field microscopy [17]. In this technique, a plasmonic nanoantenna [18,19] is introduced for achieving a sub-wavelength spatial resolution, strong field enhancement and extremely large differential cross sections of linear and non-linear optical processes (fluorescence, Raman and Rayleigh scattering, second harmonics generation, etc) [4,19,20,21]. It has become possible due to the excitation of localized surface plasmons in metallic nanostructures [5,22]. Polarization-controlled schemes are also used in photoinduced surface deformations of azobenzene polymer thin films [23,24,25] for applications in optical data storage [26] and surface relief grating formation[23].

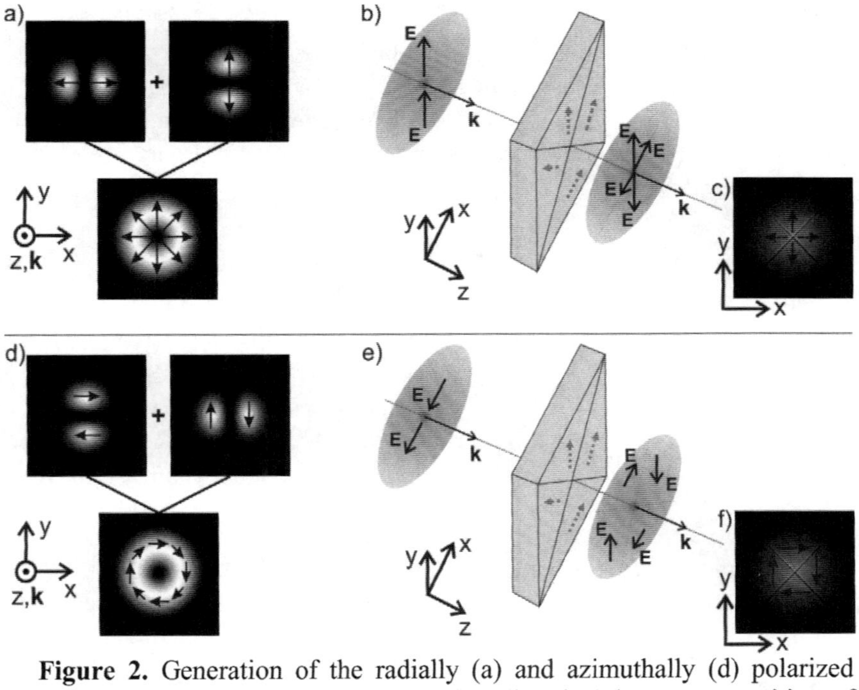


**Figure 2.** Generation of the radially (a) and azimuthally (d) polarized modes (solid arrows indicate a polarization direction) by a superposition of two Hermitte-Gaussian higher-order beams $TEM_{10}$ and $TEM_{01}$ respectively. The principle of radial (b) and azimuthal (e) mode conversion method, based on $\lambda/2$ wave plates, optical axes (depicted by dotted arrows) are oriented in appropriate direction; (c), (f) – the intensity of generated modes, calculated using FDTD (Finite-Difference Time-Domain) method.

Tightly focused higher-order laser modes were suggested for controlling the field polarization in the beam waist [4]. In particular, new states of light polarization – radially polarized (figure 2 (a)) and azimuthally polarized (figure 2 (d)) modes – appear. According to analytical calculations [4], when focusing the radially/azimuthally optical fields with a high numerical aperture lens, a strong longitudinal electric/magnetic field (up to 5 times higher compared to transverse one) arises in the focal region. Mathematically, a radially/azimuthally polarized mode is generated by a superposition of two Hermitte-Gaussian modes $\mathrm{TEM}_{10}/\mathrm{TEM}_{01}$, by rotating them by 90 degrees to each other (figure 2 (a), (d)). Although laser beams can be adjusted to a higher mode by manipulating the laser resonator, it is reasonable to convert a fundamental Gaussian beam into a higher-order mode by inserting a wave plate into different regions of the beam cross section. The advantage is that this mode conversion method does not require a perturbation of laser characteristics. Dorn et al. have demonstrated such a scheme for generation of radially and azimuthally polarized light [27]. Next, we will consider only a radial mode. As shown in figure 2 (b), a linearly polarized laser beam passes through a radial mode converter – four-section $\lambda/2$ wave plate. The optical axis of each segment is oriented in appropriate direction. Figure 2 (c) shows the FDTD (Finite-Difference Time-Domain) [28] simulation result of this mode conversion. It is obvious that the generated mode is not perfectly overlapped with the mathematical radial mode (figure 2 (a)) due to the limited number of segments, as well as the presence of additional undesirable modes. Therefore, the field structure in the waist of the highly focused radial mode generated by this approach will have differences compared to focusing of idealized (mathematical) radial mode.

In this article we provide numerical results and experimental measurements of the electric field distribution in the focal plane for the strongly focused (N.A.=1.4) radially polarized mode, generated by means of radial mode converter. The propagation of the paraxial Gaussian beam $\mathrm{TEM}_{00}$ through the radial polarizer, consisting of four $\lambda/2$ wave plates, was simulated using FDTD method. The geometry of the problem and the modeling result (generated radial mode) are shown in figure 2 (b), (c). Obtained data were used in the calculation of the field structure in the focal plane. The approach is based on angular spectrum representation and its principle depicted in figure 3. In order to determine

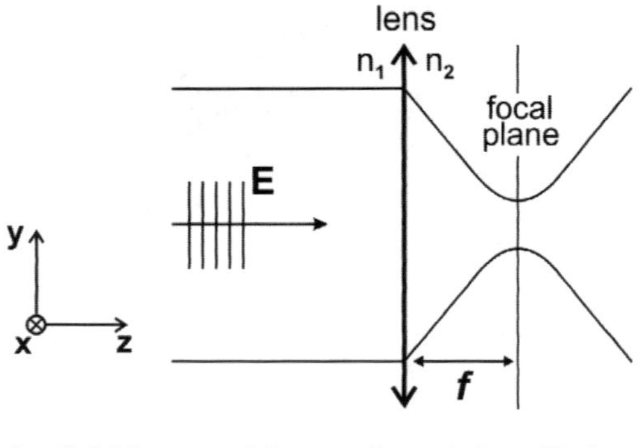


**Figure 3.** Focusing of a paraxial laser beam by aplanatic lens. The input field $\mathbf{E}$, simulated in the reference plane (coincides with the lens plane) using FDTD method, is used in calculations of the focused field. $n_1, n_2$ - refractive indices of the two media.

the field in any arbitrary plane, it is sufficient to know the field in a reference plane (simulated field $\mathbf{E}$ in this case) and the field propagator, which was established for focused light in [8,9]. Figure 4 (a) shows the calculated distribution of longitudinal and transverse components of the electric field in the focal plane. It turned out, the field distribution is no longer axially symmetric as for focused idealized radial mode (figure 4 (b)). When focusing the radially polarized mode, generated with radial mode converter, the contribution of the longitudinal field to the total field intensity is increased by a factor of $\approx 3$, as compared to linearly polarized beam. For the case of radial mode, obtained by superposition of two Hermitte-Gaussian beams $\mathrm{TEM}_{10}$, this enhancement factor is $\approx 20$. In order to verify the calculated fields, namely longitudinal field distribution, we used a modification of the azopolymer thin films produced by electric field, polarized along the propagation direction [24]. Figures 4 (d) and (e) show the photoinduced deformations of the polymer measured by atomic force microscopy (AFM).

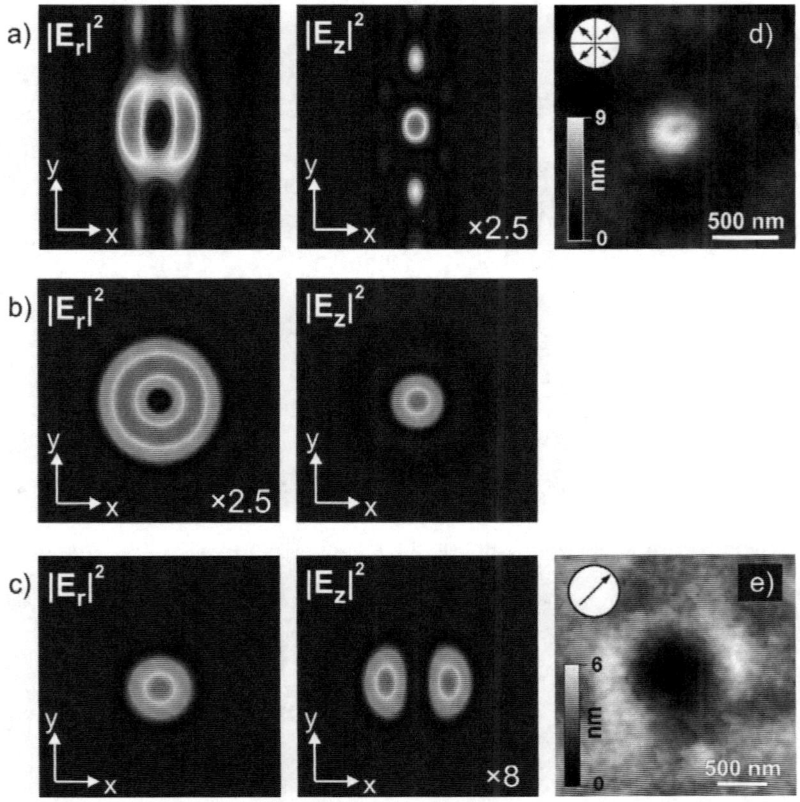


**Figure 4.** The intensity distribution of longitudinal electric field $E_z$ and transverse electric field $E_r = E_x + E_y$ in the focal plane for the highly focused (N.A.=1.4, $f_0 = 3$) radial mode, obtained with a radial mode converter (a) (image size $3\lambda \times 3\lambda$); radial mode, generated by superposition of two Hermitte-Gaussian beams (image size $3\lambda \times 3\lambda$) (b); linearly polarized in *x* direction Gaussian beam $TEM_{00}$ (c) (image size $2\lambda \times 2\lambda$). (d), (e) Photoinduced surface deformations (excitation wavelength 633 nm) of the azopolymer thin film, measured with atomic force microscopy (AFM).

The sensitivity of polymer surface reorganization was demonstrated when it was illuminated with radially polarized mode (figure 4 (a)), generated using radial mode converter, and with a linearly polarized Gaussian laser mode $TEM_{00}$ (figure 4 (c)). A photoinduced pattern replicates the calculated longitudinal component, where elevated regions are correlated with locations of large $|E_z|^2$. Finally, it should be noticed that all performed calculations are identical to an azimuthally polarized mode, which contains a longitudinal magnetic field in the vicinity of the focus.

## Acknowledgments

This work was financially supported by the Russian Foundation for Basic Research (No. 13-02-00758 A). This work was done using equipment of Federal Center of Shared Equipment of Kazan Federal University.

## References

[1] Shalaev V M 2002 *Optical Properties of Nanostructured Random Media* (Berlin: Springer)
[2] Kawata S, Ohtsu M and Irie M 2002 *Nano-optics* (Berlin: Springer)
[3] Prasad P N 2004 *Nanophotonics* (New Jersey: John Wiley & Sonsm Inc.)
[4] Novotny L and Hecht B 2006 *Principles of Nano-optics* (Cambridge: Cambridge University Press)
[5] Kawata S and Shalaev V M 2007 *Tip Enhancement Advances in Nano-Optics and Nano-Photonics* (Amsterdam: Elsevier)
[6] Celebrano M, Kukura P, Renn A and Sandoghdar V 2011 *Nature Photon.* **5** 95
[7] Yampolsky S, Fishman D A, Dey S, Hulkko E, Banik M, Potma E O and Apkarian V 2014 *Nature Photon.* DOI: 10.1038/Nhoton.2014.143
[8] Wolf E 1959 *Proc. Roy. Soc.* **253** 349
[9] Richards B and Wolf E 1959 *Proc. Roy. Soc.* **253** 358

[10] Mansipur M 1986 *J. Opt. Soc. Am.* **3** 2086
[11] Xie X S and Trautman J K 1998 *Annu. Rev. Phys. Chem.* **49** 441
[12] Sick B, Hecht B and Novotny L 2000 *Phys. Rev. Lett.* **85** 4482
[13] Novotny L, Beversluis M R, Youngworth K S and Brown T G 2001 *Phys. Rev. Lett.* **86** 5251
[14] Lu H P, Xun L and Xie X S 1998 *Science* **282** 1877
[15] Weiss S 1999 *Science* **283** 1676
[16] Fleury M, Segura J M, Zumofen G, Hecht B and Wild U P 2000 *Phys. Rev. Lett.* **84** 1148
[17] Novotny L, Sanchez E J and Xie X S 1998 *Ultramicroscopy* **71** 21
[18] Bharadwaj P, Deutsch B and Novotny L 2009 *Adv. Opt. Photon.* **1** 438
[19] Novotny L and Hulst N 2011 *Nature. Photon.* **5** 83
[20] Bharadwaj P, Beams R and Novotny L 2011 *Chem. Sci.* **2** 136
[21] Kawata S, Inouye Y and Verma P 2009 *Nature Photon.* **3** 388
[22] Maier S A 2010 *Plasmonics: Fundamentals and Applications* (Berlin: Springer)
[23] Bian S, Williams J M, Kim D Y, Li L, Balasubramanian S, Kumar J and Tripathy S 1999 *J. Appl. Phys.* **8** 4497
[24] Gilbert Y, Bachelot R, Royer P, Bouhelier A, Wiederrecht P and Novotny L 2006 *Opt. Lett.* **31** 613
[25] Kharintsev S S, Fishman A I, Kazarian S G, Gabitov I R and Salakhov M Kh 2014 *ACS Photonics* **1** 1025
[26] Rochon P, Gosselin J and Xie S 1992 *Appl. Phys. Lett.* **60** 4
[27] Dorn R, Quabis S and Leuchs G 2003 *Phys. Rev. Lett.* **91** 1
[28] Sullivan D M 2000 *Electromagnetic Simulation Using the FDTD Method* (IEEE Press)

Journal of Physics: Conference Series **613** (2015) 012011 doi:10.1088/1742-6596/613/1/012011

# Quasi-flat bands in waveguide arrays

**A I Maimistov**[1,2]

[1]Department of Solid State Physics and Nanostructures, National Research Nuclear University MEPhI, Moscow, 115409, Russia
[2]Department of Physics and Technology of Nanostructures, Moscow Institute for Physics and Technology, Dolgoprudny, Moscow region, 141700 Russia.

E-mail: aimaimistov@gmail.com

**Abstract**. Coupled electromagnetic waves propagating in a waveguide array are discussed. This waveguide array can be considered as a one dimensional photon crystal composed from the unit cell containing three different waveguides, namely, two positive refraction index waveguides and one waveguide with negative refraction index. Spectrum of the linear modes of electromagnetic waves in this device contains the quasi-flat band.

## 1. Introduction

In the last twenty – twenty-five years new materials for optics have been fabricated. The best known example for these materials is the photon crystal or the photonic band-gap material [1] and metamatrerials with negative refractive index. The use of these materials allows investigating different physics phenomena by optical simulation methods. For example, the optical simulation of Majorana physics was considered in [2] via two one-dimensional photonic crystals. In [3] the photonic realization of the quasi-particles carrying fractional statistics that interpolate between bosons and fermions was proposed. There are series of works devoted to photonic crystal waveguide analogy of the graphene [4, 5] and to the realization of the Dirac point with double cones in optics [6-8]. Review of the optical analogies using photonic structures is presented in [9].

In the case of 2D photonic crystals under some conditions there is an energy band that is pinned to the Dirac points [10-14]. This band can be made very flat by controlling the parameters of the photonic crystals. It should be noted that the flat band (or flat zone) is formed if the photonic crystal unit cell contains three different "atoms". In the case under consideration "atom" is waveguide. Here waveguide array as 1D photonic crystal will be considered.

## 2. Alternating optical waveguide arrays

Two closely located waveguides can be coupled due to the tunneling of light from one waveguide to the other. This phenomenon forms the basis for the device which is named coupler. A coupler using waveguides fabricated from materials with a positive refractive index (PRI) preserves the direction of light propagation, and for this reason it is named a directional coupler. If one of the waveguides of the coupler is fabricated from a material with a negative refractive index, then a radiation entering one waveguide leaves the coupler through the other waveguide at the same end but in the opposite direction [15]. For this reason this device was called the oppositely directional coupler.

Let us consider the waveguide array where the waveguide having number $n$ is characterized by positive refractive index, whereas the nearest neighbouring waveguides with numbers $n-1$ and $n+1$ possess negative refractive index (NRI). If the electromagnetic radiation is localized in each waveguide the coupled wave theory can be used. The configuration of these alternating waveguides will be marked as alternating optical waveguide arrays (AOWA). The dispersion relation for a linear wave in AOWA consists of two branch separated by a forbidden zone (energy gap). It is important to emphasize that the unit cell of AOWA contains two different waveguides, figure 1. In this figure $A_n$ ($B_n$) is a slowly varying complex envelope of electric field in the PRI-waveguide (NRI-waveguide) possessed by the $n$th unit cell.

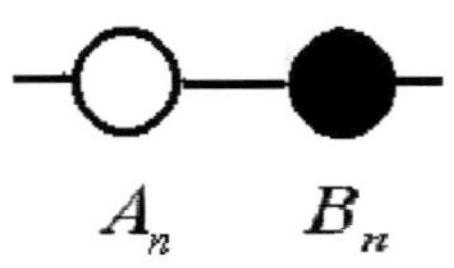


**Figure 1.** Unit cell of alternating optical waveguide arrays.

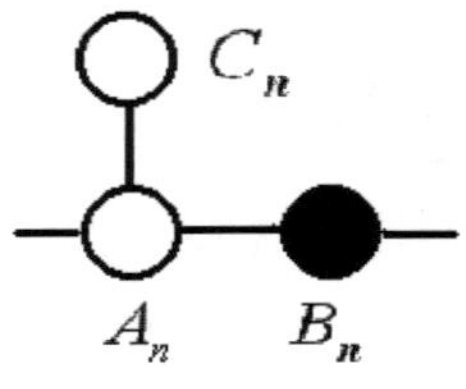


**Figure 2.** Unit cell of modified alternating optical waveguide arrays.

Let us suppose that the waveguide array is composed from the unit cell containing three waveguides: two PRI-waveguides and one NRI-waveguide (figure 2). Here $A_n$ and $C_n$ are slowly varying complex envelopes of the electric field in the PRI-waveguides. The $B_n$ is a slowly varying complex envelope of electric field in the NRI-waveguide. The system of the normalized equations describing electromagnetic wave propagation in the this linear structure without taking into account the group velocity dispersion reads

$$i(A_{n,t}+A_{n,x})+B_{n-1}+B_n+\beta C_n=0, \quad (1)$$

$$i(B_{n,t}-B_{n,x})+A_n+A_{n+1}=0, \quad (2)$$

$$i(C_{n,t}+C_{n,x})+\beta A_n=0, \quad (3)$$

It is suggested that the group velocities and the propagation constants are equal for all channels. The coordinate $x$ is a ration of the distance along the waveguide to the coupling distance between the PRI-waveguide and the NRI-waveguide. The coefficient $\beta$ is a ration of the coupling constant between the neighbor PRI-waveguides to the coupling constant between the PRI-waveguide and NRI-waveguide.

Let us suppose that the waveguide array is rolled up to ring. It results in the following boundary conditions $A_{N+1}=A_1$, $B_{N+1}=B_1$ and $C_{N+1}=C_1$, where $N$ is the number of the ring cells. If the following ansatz

$$A_n(x,t)=a_n e^{-i\omega t+ikx}, \quad B_n(x,t)=b_n e^{-i\omega t+ikx}, \quad C_n(x,t)=c_n e^{-i\omega t+ikx},$$

is used, then the system of equations (1)-(3) reduces to equations

$$(\omega-k)a_n+b_{n-1}+b_n+\beta c_n=0, \quad (\omega+k)b_n+a_n+a_{n+1}=0, \quad (\omega-k)c_n+\beta a_n=0. \quad (4)$$

This system of algebraic equations can be transformed to diagonal form by using the discrete Fourier transformation.

$$a_n = \sum_{s=-M}^{s=M} \tilde{a}_s \exp\{2\pi i sn/N\}, \quad b_n = \sum_{s=-M}^{s=M} \tilde{b}_s \exp\{2\pi i sn/N\}, \quad c_n = \sum_{s=-M}^{s=M} \tilde{c}_s \exp\{2\pi i sn/N\},$$

where $M = N/2$ or $M = (N-1)/2$ is an integer number. Substitution of this expression into (4) results in the following system of equations

$$(\omega - k)\tilde{a}_s + z\tilde{b}_s + \beta\tilde{c}_s = 0, \quad (\omega + k)\tilde{b}_s + z^*\tilde{a}_s = 0, \quad (\omega - k)\tilde{c}_s + \beta\tilde{a}_s = 0, \tag{5}$$

where $z = 1 + \exp(-2\pi i s/N)$. System of equations (5) has a nonzero solution only if the corresponding determinant is equal to zero. That leads to implicit form of the dispersion relation

$$(\omega - k)[(\omega^2 - k^2) - |z|^2] - \beta^2(\omega + k) = 0. \tag{6}$$

At $k = 0$ from (6) tree values of $\omega_s(0)$ follow: $\omega_s^{(1,2)}(0) = \pm(\beta^2 + |z|^2)^{1/2}$ and $\omega_s^{(0)}(0) = 0$. Hence three branches of the dispersion relation are present at each value of $s = 0, \pm 1, \pm 2, ..., \pm M$.

Figures 3 and 4 show several examples of the dispersion curves at $s = 0$.

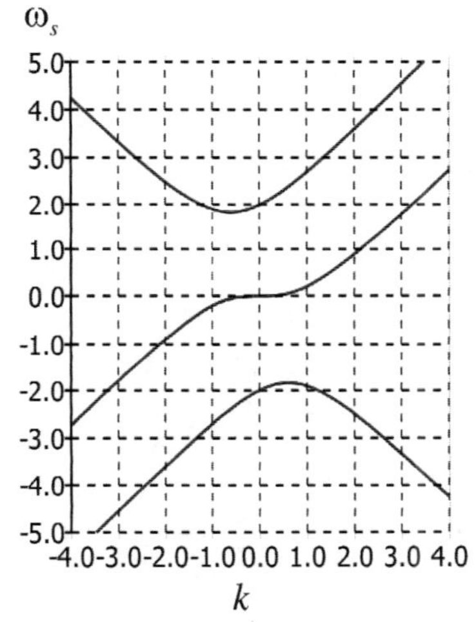


**Figure 3.** Dispersion curves at $|z|^2 = 2.0$ $\beta^2 = 1{,}98$.

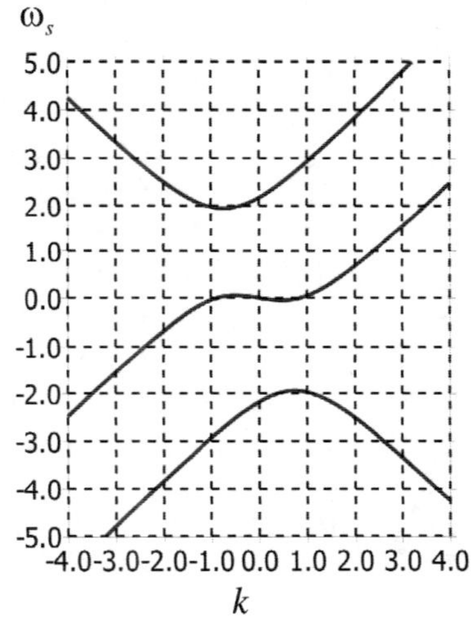


**Figure 4** Dispersion curves at $|z|^2 = 2.0$ $\beta^2 = 2{,}75$.

If the coupling with the third waveguide (i.e., where the electric field is denoted as $C_n$) is equal to zero, then dispersion equation (6) leads to the famous result [16] for modes with gap

$$\omega_s^2(k) = |z|^2 + k^2,$$

and gapless mode $\omega_s(k) = k$.

Near the point ($k = 0, \omega = 0$) dispersion equation (6) can be solved approximately. This leads to the dispersion relation for the gapless mode

$$\omega_s(k) \approx \frac{|z|^2 - \beta^2}{|z|^2 + \beta^2} k. \tag{7}$$

From equation (7) it follows that the wave associated with the gapless mode is akin to light wave in vacuum. However, the velocity of this wave can be very nearly equal to zero.

It is interesting to observe that in the case of an infinite waveguide array (i.e., $N \to \infty$), at $k = 0$ the dispersion relations are presented by the following expressions

$$\omega^{(\pm)}(0,q) = \pm(\beta^2 + 4\cos^2 ql/2)^{1/2} \quad \omega^{(0)}(0,q) = 0,$$

where $q$ is the transverse wave number, $l$ is the lattice constant. It follows from this expressions that the spectrum of the gapless mode belongs to the flat band of 1D photon crystal. Because the true dispersion relation for the gapless mode in not plane, we can say about quasi-flat band only.

## 3. Acknowledgments

I would like to thank Dr. A.S. Desyatnikov, Dr. E.V. Kazantseva and A.O. Karavaeva for interesting and helpful discussions. The research was supported by Russian Science Foundation (project 14-22-00098).

## References

[1] Yablonovitch E 1994 *J. Mod. Opt.* **41** 173
[2] Rodri'guez-Lara B M and Moya-Cessa H M 2014 *Phys. Rev.* A **89** 015803
[3] Longhi S and Valle G D 2012 *Opt. Lett* **37** 2160
[4] Benisty H 2009 *Phys. Rev.* B **79** 155409
[5] Mihalache I and Dragoman D 2011 *J. Opt. Soc. Amer.* B **28** 1746
[6] Sakoda K 2006 *J. Opt. Soc. Amer.* B **29** 2770
[7] Wang Li-Gang, Wang Zhi-Guo, Zhang Jun-Xiang and Zhu Shi-Yao 2009 *Optics Letters* **34** 1510
[8] Shen Ming, Ruan Lin-Xu and Chen Xi 2010 *Optics Express* **18** 12779
[9] Longhi S 2009 *Laser & Photonics Rev.* **3** 243
[10] Apaja V, Hyrka"s M and Manninen M 2010 *Phys. Rev.* A **82** 041402(R)
[11] Jukic D, Buljan H, Lee D-H, Joannopoulos J D and Soljacic M 2012 *Optics Letters* **37** 5262
[12] Chui S T, Liu Shiyang and Lin Zhifang 2013 *Phys. Rev.* E **88** 031201(R)
[13] Chern Gia-Wei, Chien Chih-Chun and Di Ventra M 2014 *Phys. Rev.* A **90** 013609
[14] Leykam D, Bahat-Treidel O and Desyatnikov A S 2012 *Phys. Rev.* A **86** 031805(R)
[15] Litchinitser N M, Gabitov I R and Maimistov A I 2007 *Phys. Rev. Lett.* **99** 113902
[16] Kazantseva E V, Maimistov A I and Ozhenko S S 2009 *Phys. Rev.* A **80** 43833

Journal of Physics: Conference Series **613** (2015) 012012 doi:10.1088/1742-6596/613/1/012012

# On solitons in the negative refracting medium

**A I Maimistov**[1,2], **I R Gabitov**[3] **and E I Lyashko**[2]

[1]Department of Solid State Physics and Nanostructures, National Research Nuclear University MEPhI, Moscow, 115409, Russia
[2]Department of Physics and Technology of Nanostructures, Moscow Institute for Physics and Technology, Dolgoprudny, Moscow region, 141700 Russia
[3] Department of Mathematics, University of Arizona, Tucson, Arizona 85721, USA

E-mail: aimaimistov@gmail.com

**Abstract**. We consider the coupled electromagnetic waves propagating in a nonlinear medium, which is featured by a positive and negative refraction indexes. The backward waves can be propagating in this case. The example of the true soliton is discussed. In general case the coupled forward and backward solitary wave can be found. They are analogues to the optical solitons.

## 1. Introductionn

As a rule, the phase velocity ($\boldsymbol{V}_{\rm ph}$) and the Pointing vector ($\boldsymbol{S}$) of the electromagnetic wave are collinear vectors. Frequently the Pointing vector is considered to be proportional to the group velocity vector. It is correct in the case of isotropic medium. Antiparallel orientation of the phase velocity and the Pointing vector (or group velocity vector) was first discussed in [1, 2]. In [3] it was indicated that antiparallel orientation of $\boldsymbol{V}_{\rm ph}$ and $\boldsymbol{S}$ results in negative refraction. Subsequently, this idea was developed by Mandelstam in [4]. It has been predicted that when the real parts of the dielectric permittivity and magnetic permeability in the medium simultaneously take on negative values in some frequency range, antiparallel orientation of the $\boldsymbol{V}_{\rm ph}$ and $\boldsymbol{S}$ occurs [5,6] and the property of negative refraction appears [7]. The existence of the media characterized by negative refractive index (NRI) was demonstrated experimentally first in the microwave and then in the near-infrared ranges. Reviews of the properties of the NRI materials and the nonlinear phenomena in NRI materials are presented in [8-13].

## 2. Sine-Gordon equation. Forward and backward solitons

The Sine-Gordon equation is the famous example of the evolution equation in the nonlinear wave theory [14, 15]. This equation is derived in many fields of physics and its properties are well known. In the one dimension case the Sine-Gordon equation is completely integrable one. Solitons and multi-solitons (in the strict sense) are the solutions of this equation. Thus the Sine-Gordon equation can be considered as an etalon equation in this context.

There are two forms of the Sine-Gordon equation

$$u_{,xx} - u_{,tt} = a\sin u\,, \qquad (1)$$

and

$$u_{,xt} = a \sin u . \tag{2}$$

Linearization of these equations results in linear wave equations for description of the linear wave propagation

$$u_{,xx} - u_{,tt} - au = 0, \tag{3}$$

$$u_{,xt} - au = 0 . \tag{4}$$

From equation (3) the dispersion relation follows in the form $\omega^2(k) = a + k^2$. The phase and group velocities are determined by the following expression

$$V_{ph} = \frac{\omega}{k} = \pm \frac{(a+k^2)^{1/2}}{k}, \quad V_g = \frac{d\omega}{dk} = \pm \frac{k}{(a+k^2)^{1/2}} .$$

It is necessary to emphasize that signs of the phase and group velocities are the same. Hence, the linear waves in this case are the forward waves.

Equation (4) results in the dispersion relation in the form $\omega(k) = ak^{-1}$. Thus, the phase and group velocities are determined by the following expression

$$V_{ph} = \frac{\omega}{k} = \frac{a}{k^2}, \quad V_g = \frac{d\omega}{dk} = -\frac{a}{k^2} .$$

In this case the signs of the phase and group velocities are opposite ones. Hence, the linear waves in this case are the backward waves.

If we are supposed that equation (1) is the nonlinear generalization of equation (3), we can say equation (1) describes the forward wave soliton. Oppositely, if the Sine Gordon equation (2) is the nonlinear generalization of equation (4), the solitons to equation (2) can be treated as the backward wave solitons.

Another example of the backward wave solitons has been represented in [16], where backward wave propagation was considered by using the Korteweg-de Vries equation. Propagation of the backward solitary waves (they are not solitons in the strict sense) was discussed in many papers. Review of the nonlinear phenomena in this field is presented in [12].

## 3. Coupled forward and backward waves

New phenomena in nonlinear optics were predicted in the case of the forward and backward wave interaction. Second and third harmonic generation in NRI metamaterials can be accompanied by quadratic and cubic backward wave soliton formation [17-20]. The forward wave and backward wave interaction is realized in a nonlinear coupler and in a nonlinear waveguide array, which consists of alternating waveguides of positive and negative refraction indexes [21-24]. Gap solitons and the bistability of continuous waves in an oppositely directed coupler represent new effects due to the positive-negative refraction phenomenon [21, 24].

The characteristic form of the evolution equations describing the forward and backward wave interaction is

$$i(e_{1,x} + v_1^{-1} e_{1,t}) = P_1[e_1, e_2], \quad i(-e_{2,x} + v_2^{-1} e_{2,t}) = P_2[e_1, e_2], \tag{5}$$

where $e_1$ ($e_2$) is a slowly varying envelope of forward (backward) wave, $v_1$ ($v_2$) is the group velocity of forward (backward) wave, and $P_1$ ($P_2$) is a functional describing the dispersion and coupling effects. For example, in the second harmonic generation case these functional can be written as

$$P_1[e_1, e_2] = -s_1 e_{1,tt} + e_2 e_1^*, \quad P_2[e_1, e_2] = -s_2 e_{2,tt} + e_1^2 .$$

At this point the forward and backward wave propagating in cubic nonlinear medium will be considered. Let us assume that thephase velocities are collinear but the group velocities are oppositely directed. The normalized system of evolution equations can be written in following form

$$i(e_{1,x} + v_1^{-1} e_{1,t}) = s_1 e_{1,tt} + (|e_1|^2 + |e_2|^2) e_1, \quad (6)$$

$$i(-e_{2,x} + v_2^{-1} e_{2,t}) = s_2 e_{2,tt} + (|e_1|^2 + |e_2|^2) e_2, \quad (7)$$

where $s_1$ ($s_2$) is the coefficient of group velocity dispersion for the forward (backward) wave.

If the following ansatz

$$e_1(y) = a_1(y) e^{i\omega_1 t + i\kappa_1 x}, \quad e_2(y) = a_2(y) e^{i\omega_2 t + i\kappa_2 x}, \quad y = \tau_p^{-1}(t - x/V + t_0),$$

is used, then the family of steady state solutions of the system of equations (6)-(7) can be found. Here $\omega_{1.2}$, $\kappa_{1,2}$, $\tau_p$ and $t_0$ are parameters of normalization. Parameter $V$ is the solitary wave velocity.

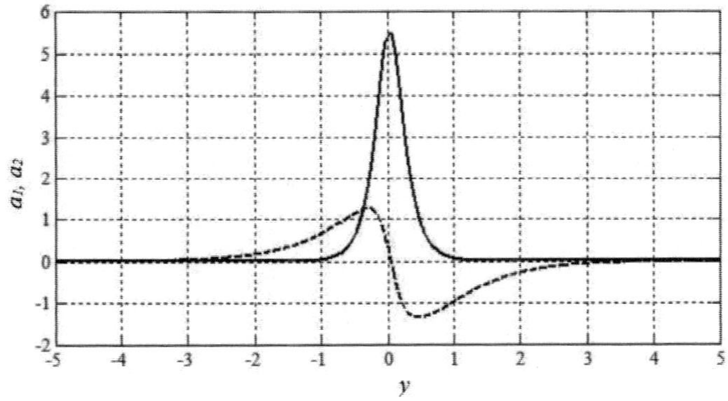


**Figure 1**. Real envelope of the forward soliton (full line) and backward soliton (dashed line) as function of characteristic variable $y$.

The kind of steady state solutions of system of equations (6)-(7) is defined by the boundary conditions at infinity ($y \to \pm\infty$). Thus, there are bright-bright solitary waves (i. e., a coupled bright forward wave and a bright backward wave) and dark-bright solitary waves (for example, a coupled bright forward wave and a dark backward wave). The first kind of the solitary wave is shown in figure 1. At some condition on the parameters $\omega_{1.2}$, $\kappa_{1,2}$ and $v_{1,2}$ the amplitudes $a_1$ and $a_2$ can be the same.

## 4. Acknowledgments

The research was supported by Russian Science Foundation (project 14-22-00098). I.R. Gabitov was partially supported by NSF grant DMS-0509589, ARO-MURI award 50342-PH-MUR.

## References

[1] Lamb H 1904 *Proc Proc. London Math. Soc.* **1** 473
[2] Pocklington H C 1905 *Nature* **71** 607
[3] Schuster A 1904 *An Introduction to the Theory of Optics* (London: Edward Arnold)
[4] Mandel'shtam L I 1945 *Zh. Eksp. Teor. Fiz.* **15** 475

[5] Sivukhin D V 1957 *Opt. Spektrosk.* **3** 308
[6] Pafomov V E 1959 *Sov.Phys. JETP* **9** 1321
[7] Veselago V G 1968 *Sov. Phys. Usp.* **10** 509
[8] Veselago V, Braginsky L, Shklover V and Hafner Ch 2006 *J. Comput. and Theor. Nanoscience.* **3** 189
[9] Agranovich V M and Gartstein Yu N 2006 *Phys. Usp.* **49** 1029
[10] Litchinitser N M, Gabitov I R, Maimistov A I and Shalaev V M 2008 Negative refractive index metamaterials in optics *Progress in Optics* vol. 51 ed. E Wolf. (Amsterdam: Elsevier) chapter 1 pp. 1-68
[11] Maimistov A I and Gabitov I R 2007 *Eur. Phys. J. Special Topics* **147** 265
[12] Lapine M, Shadrivov I V and Kivshar Yu S 2014 *Rev. Mod. Phys.* **86** 1093
[13] Maimistov A I and Gabitov I R 2015 Nonlinear Optical Effects in Positive-Negative Refractive Index Materials. *Nonlinear, Tunable and Active Metamaterials* ed. I V Shadrivov, M Lapine and Yu S Kivshar (Springer-Verlag. Berlin) pp 133–158
[14] Barone A, Esposito F, Magee C J and Scott A C 1971 *Riv.Nuovo Cimento* A **1** 227
[15] Scott A. 2003 *Nonlinear Science: Emergence and Dynamics of Coherent Structures*, 2nd ed. (Oxford: OUP).
[16] Weina Cui, Yongyuan Zhu, Hongxia Li and Sumei Liu 2010 *Phys.Lett.* A **374** 380
[17] Maimistov A I, Gabitov I R and Kazantseva E V *Optics and Spectroscopy* **102** 90
[18] Roppo V, Centini M, Sibilia C, Bertolotti M, de Ceglia D, Scalora M, Akozbek N, Bloemer M J, Haus J. W, Kosareva O G and Kandidov V P 2007 *Phys.Rev.* A **76** 033829 (12 pages)
[19] Roppo V, Ciraci C, Cojocaru C and Scalora M 2010 *J.Opt.Soc.Amer.* B **27** 1671
[20] Elyutin S O, Maimistov A I and Gabitov I R 2010 *JETP* **111** 157
[21] Litchinitser N M, Gabitov I R and Maimistov A I 2007 *Phys. Rev. Lett.* **99** 113902
[22] Kazantseva E V, Maimistov A I and Ozhenko S S 2009 *Phys.Rev.* A **80** 43833 (7 pages)
[23] Kazantseva E V and Maimistov A I 2013 *Quantum Electron.* **43** 807
[24] Ryzhov M S and Maimistov A I 2012 *Quantum Electron.* **42** 1034

# New possibilities of photon echo: determination of ground and excited states g-factors applying a weak magnetic field pulse

**V N Lisin[1], A M Shegeda[1] and V V Samartsev[2]**

[1]Zavoisky Physical-Technical Institute, Russian Academy of Sciences, Kazan, 420029 Russia
[2]Kazan (Volga region) Federal University, Kazan, 420008, Russia

E-mail: vlisin@kfti.knc.ru

**Abstract**. We have measured the magnetic parameters of a paramagnetic ion in the ground and excited states by controlling the relative phases of excited dipoles with pulse of a weak magnetic field. $Er^{3+}$ in two crystals $LuLiF_4$ and $YLiF_4$ has been used as a paramagnetic ion. The second matrix was used for control. Optical transition is $^4I_{15/2} \Rightarrow 4F_{9/2}$.

## 1. Introduction

A weak perturbation pulse can lead to a significant change of the relative phases of radiators and so to the amplitude of the resulting dipole moment. It can be manifest in many cases, in which coherent effects are important (photon echo, the propagation of light, and so forth). In the optical systems, in which the Zeeman Effect is present, a pulse magnetic field can be used as a perturbation. The Zeeman Effect is the change of the period of light emitted by a substance when the substance was acted upon by magnetic force [1]. Pieter Zeeman observed only the broadening of optical lines in a magnetic field but not splitting. The frequency change of the emitted light should exceed the inhomogeneous line width, in order to observe the splitting of the optical line. A large magnetic field is needed to use this. Photon echo is the method which overcomes the difficulties associated with inhomogeneous line width and allows to observe the splitting of optical lines in a weak magnetic field.

## 2. Photon echo

Photon echo is coherent radiation of a system in the form of the short pulse, caused by restoration of the phase of separate radiators after the change of a sign on relative frequency of radiators. For two-pulse photon echo we have

$$\begin{gathered}\phi_\alpha(t) = \delta_\alpha(t_2 - t_1) + \pi - \delta_\alpha(t - t_2) = \{t = t_{echo} = 2t_2 - t_1\} = \pi, \\ P(t) \propto \sum_\alpha e^{i\phi_\alpha(t)} g(\delta_\alpha) = \{t = t_{echo} = 2t_2 - t_1\} = -P(0), \\ I(t) = |P(t)|^2 = \{t = t_{echo} = 2t_2 - t_1\} = I(0), \quad \delta_\alpha = (\omega_\alpha - \omega).\end{gathered} \tag{1}$$

Here $t_1$ and $t_2$ are the times of switching on laser pulses, $\omega$ is the laser frequency, $\omega_\alpha$ is frequency of transition for α-ion, $P$ is the dipole moment of a system, $I$ is echo intensity.

### 3. Zeeman splitting of optical line by magnetic field

Transition frequencies of two groups of paramagnetic ions $Er^{3+}$ with spin projection in ground state ±1/2 are shifted by magnetic field on the different values (see figure 1)

$$\begin{aligned}&\Delta E\left({}^4\mathrm{I}_{15/2}\right)=\beta H\langle\mu=\pm 3/2|L_z+2S_z|\mu=\pm 3/2\rangle=\\&=g\left({}^4\mathrm{I}_{15/2}\right)\beta Hm,\ m=\pm 1/2\\&\Delta\omega=Zm,\ \ (rad/s)\\&Z=\left[g({}^4\mathrm{F}_{9/2})\pm g({}^4\mathrm{I}_{15/2})\right]\beta H/\hbar,\end{aligned}\tag{2}$$

where $g({}^4\mathrm{F}_{9/2})$ and $g({}^4\mathrm{I}_{15/2})$ are the g-factors of excited and ground states of erbium ion, β is the Bohr magneton, $\hbar$ is the Planck constant, $\mu$ is crystal quantum number, $L_z$ an $S_z$ are projections on crystal axis $C$ of total orbital and spin moments accordingly for $Er^{3+}$. You can see that $\pi$- and $\sigma$-Zeeman splitting is proportional accordingly to the difference and to the sum of the excited and ground states g-factors.

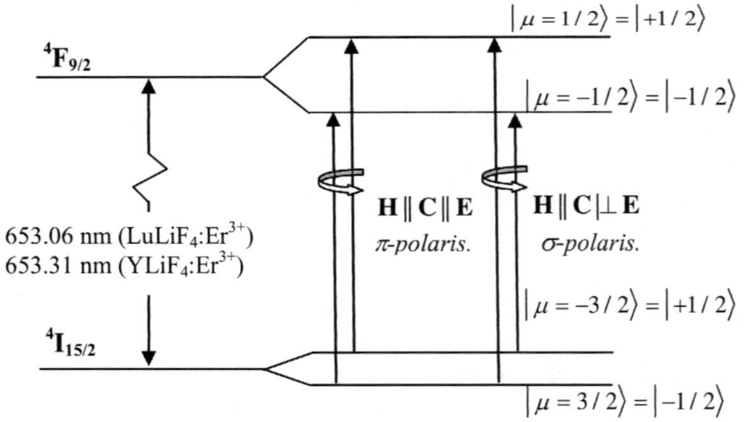


**Figure 1.** Energy levels of $Er^{3+}$ in $YLiF_4$ and $LuLiF_4$ (**H**‖**C**). Also the transition and the transition wavelength are shown (the arrows indicate the transitions for $\pi$- and σ- polarizations of the laser). **C** is crystal axes, **H** is magnetic field and E is laser electrical field. **E**‖**C** is $\pi$- polarization, **E**⊥**C** is $\sigma$-polarization. $\mu$ is crystal quantum number, $|\pm 1/2\rangle$ indicate the $Er^{3+}$ ion states whose spin projection on axis C are ±1/2.

### 4. Photon echo and a pulse of magnetic field

If the magnetic field pulse switches on after the second laser pulse then the relative phase $\varphi$ of two groups of radiators is not equal to zero:

$$\begin{aligned}&\phi_\alpha(t)=\delta_\alpha(t_2-t_1)+\pi-\delta_\alpha(t-t_2)-\int_{t0}^{t}dt'\Delta\omega(t')=\phi_\alpha{}^0(t)-m_\alpha\varphi(t),\\&\varphi(t)=\int_{t0}^{t}Z(t'),\ \ m_\alpha=\pm 1/2,\end{aligned}\tag{3}$$

and the amplitude of the total dipole moment and the echo intensity are significantly changed:

$$P(t) \propto \sum_{\alpha} e^{i\phi_\alpha(t)} g(\delta_\alpha) = \sum_{\alpha} e^{i\phi_\alpha{}^0(t)} e^{-im_\alpha\varphi(t)} g(\delta_\alpha) = P_0(t)\cos(\varphi(t)/2),$$
$$I(t) = |P(t)|^2 = I_0(t)(1+\cos(\varphi(t)))/2, \tag{4}$$

where $t_0$ is the time of the magnetic pulse switch on.

## 5. Magnetic pulse does not overlap in time with echo-pulse

The relative phase $\varphi$ is not dependent on time in this case. The echo intensity oscillates versus relative phase as you can see from (4) and figure 2. Zeeman splitting Z=10 MHz was observed [2] in $YLiF_4:Er^{3+}$ and $LuLiF_4:Er^{3+}$, which is much less than the inhomogeneous line width =600 MHz.

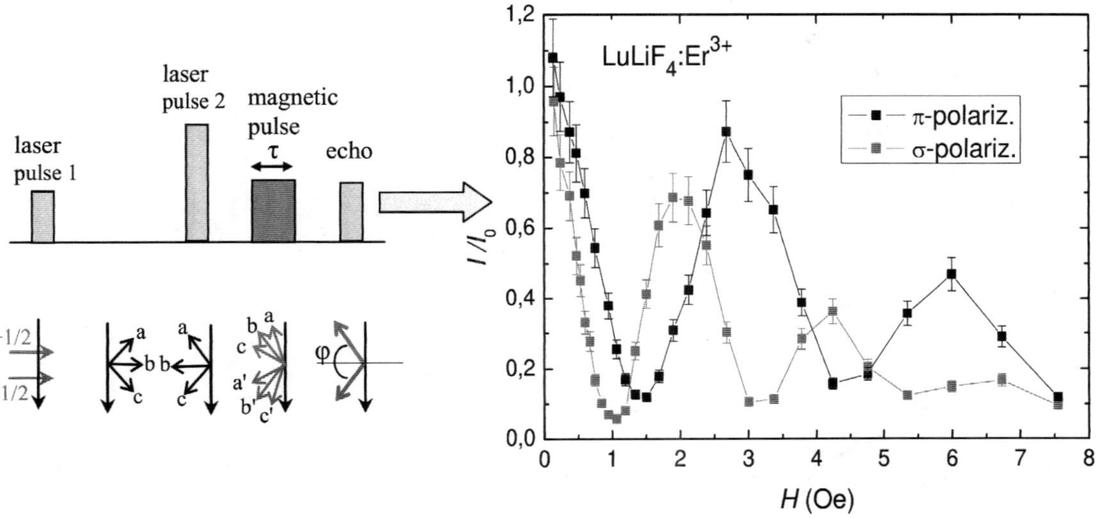


**Figure 2.** Time sequence of laser, echo and magnetic pulses, phase's evolution (left) and relative echo intensity versus magnetic pulse amplitude for sigma- and pi- polarizations in $LuLiF_4:Er^{3+}$.

## 6. Magnetic pulse overlaps in time with echo-pulse

The phase $\varphi$ depends on time. The echo time form is changed [3]. If the relative phase $\varphi < \pi$ then the echo pulse is "compressed" and the time of echo "decreases". It is shown in figure 3 (above): schematically on the left and experimental results on the right. .If the relative phase $\varphi$ is larger than $3\pi$ then the echo time form is modulated. Figure 3 (below) shows the case of $\varphi > 5\pi$: schematically on the left and experimental results on the right. The modulation period $T$ is determined from the relation

$$\varphi(t+T(t)) - \varphi(t) = 2\pi \ . \tag{5}$$

It generally depends on time. If echo $I_0$ and Zeeman splitting $Z$ are constant in the time interval $[t, T(t)+t]$) then (see (3))

$$\varphi(t+T(t)) - \varphi(t) = 2\pi = \int_t^{t+T(t)} Z(t')dt' \approx Z(t)T(t) \,. \tag{6}$$

In our experiment, as you can see in figure 3 (below), the duration of the flat part on top of the magnetic pulse (25 ns) becomes approximately equal to the echo pulse duration (28 ns). Therefore, $Z$

is constant in (6). From the waveforms of echo signals observed at various times $t_0$ of the beginning of action of the magnetic pulse, similar to those shown in figure 3, the time intervals between the nearest minima in the echo signals were measured and the average value $T$ of the modulation period was determined. The next equation has been used for the determination of g-factors:

$$T\left[g(^4\mathrm{F}_{9/2}) \pm g(^4\mathrm{I}_{15/2})\right]\beta H / \hbar = 2\pi \ . \tag{7}$$

Here $H$ is the magnetic pulse amplitude in the region of the flat part on top of the pulse.

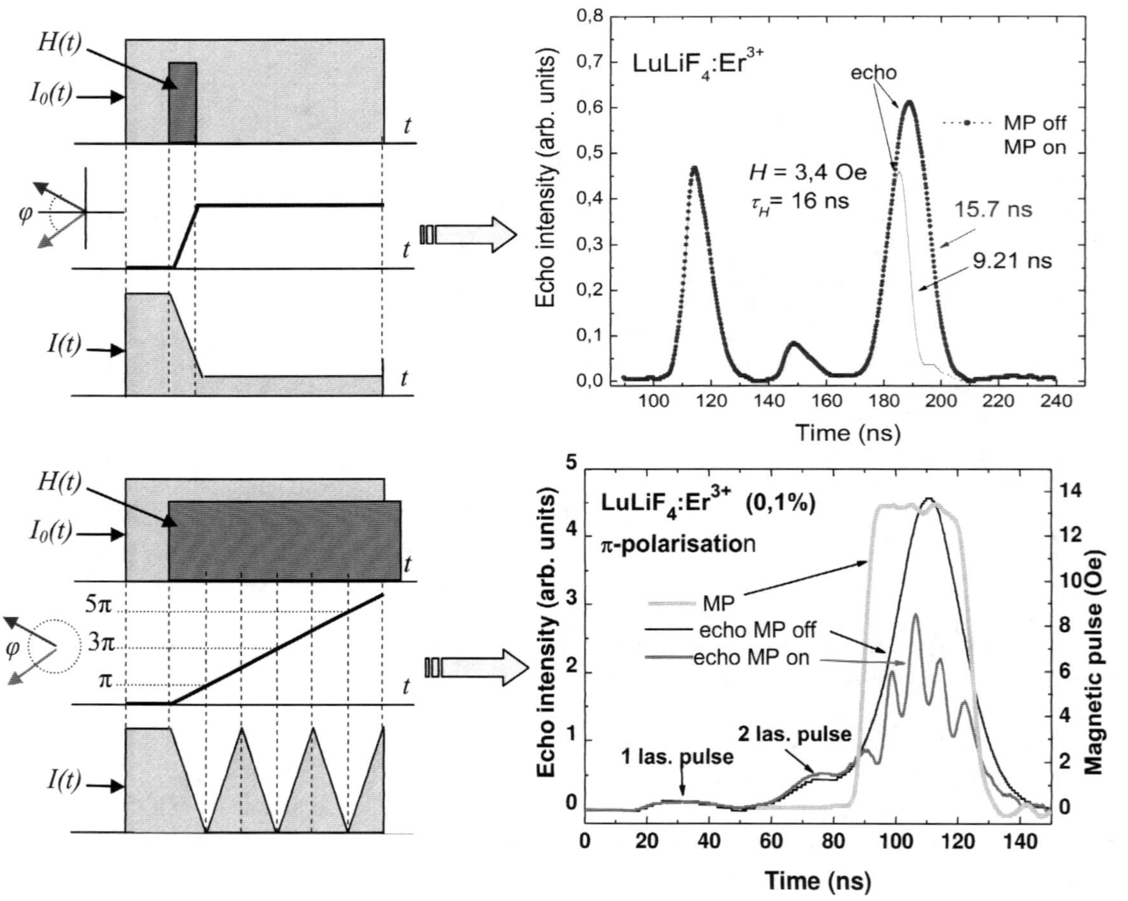


**Figure 3.** Magnetic pulse overlaps in time with echo-pulse. The case $\varphi < \pi$ is shown above and $\varphi > 5\pi$ is shown below. It illustrates the waveforms of magnetic pulse and photon echo for $\pi$-polarizations of laser pulses when the magnetic pulse is on and off.

The dependence of the average period $T$ versus $H$ has been built. Table 1 shows these dependences for different polarizations of the laser pulses in the samples $YLiF_4:Er^{3+}$ and $LuLiF_4:Er^{3+}$. Also the values of the sum and differences of g-factors, determined using (7), are shown.

**Table 1.** $T$ versus the $H$ for different polarizations of the laser pulses

| | | | | | | | | | |
|---|---|---|---|---|---|---|---|---|---|
| LuLiF4:Er3+ | $H$, Oe | 19.44 | 17.17 | 15.08 | 13.3 | 11.74 | 10.42 | 9.19 | 8.15 |
| $\pi$-polariz. | $T$, ns | 5.61 | 6.38 | 7.25 | 8.25 | 9.27 | 10.39 | | |
| | $g'-g$ | 6.55 | 6.52 | 6.53 | 6.51 | 6.56 | 6.6 | | |
| $\sigma$- polariz. | $T$, ns | | | | | | 5.4 | 6.15 | 6.95 |
| | $g'+g$ | | | | | | 12.72 | 12.64 | 12.61 |
| YLiF4:Er3+ | $H$, Oe | 13.45 | 11.87 | 10.44 | 9.2 | 8.11 | 7.21 | 6.36 | 5.63 |
| $\pi$- polariz. | $T$, ns | | 9.11 | 10.25 | 11.78 | | | | |
| | $g'-g$ | | 6.61 | 6.68 | 6.59 | | | | |
| $\sigma$- polariz. | $T$, ns | | | | | 6.7 | 7.76 | 8.76 | 9.95 |
| | $g'+g$ | | | | | 13.14 | 12.77 | 12.82 | 12.75 |

Here $g' = g(^4F_{9/2})$ and $g = g(^4I_{15/2})$ are g-factors of the excited and ground states of $Er^{3+}$.

Determining from this table the average value of the sum and the difference of the g-factors we find

$$g = 3.06, \quad g' = 9.6 \quad \text{for } LuLiF_4 : Er^{3+}$$

$$g = 3.12, \quad g' = 9.75 \quad \text{for } YLiF_4 : Er^{3+}$$

The ground state g-factor values are in good agreement with those measured by EPR method: $g(LuLiF_4{:}Er^{3+})=3.09$ [4], $g(YLiF_4{:}Er^{3+})=3.14$ [5]. The excited state g-factor value for $YLiF_4{:}Er^{3+}$ is in good agreement with those measured by optical methods: $g'(YLiF_4{:}Er^{3+}) = 9.84 \pm 0.25$ [6], $g'(YLiF_4{:}Er^{3+}) = 9.61$ [7].

## 7. Conclusion

In our study we measured the Zeeman splitting of the ground and excited states and determined magnetic parameters of the paramagnetic ion with an accuracy comparable to the EPR using the known values of the magnetic pulse amplitude. If the magnetic parameters are known, it is possible to determine the amplitude of the MP, using the relation (7). For example, if the magnetic field is created by the pulse change of the dipole-dipole interaction of the ion with the environment. This can be used to determine the distance to the centers of the environment by measuring the waveform of the echo modulation periods, not only in the optical range, but also in the EPR and NMR bands, using relations similar to (7).

## Acknowledgments

This research has been supported partially by the Russian Foundation for Basic Research (projects no. 14-02-00041-a and no. 14-02-90000Bel-a) and the Division of Physical Sciences, Russian Academy of Sciences (programs “Fundamental Optical Spectroscopy and its applications”.

## References

[1] Zeeman P 1897 *Nature* **55** 347-347
[2] Lisin V, Shegeda A and Gerasimov K 2012 *JETP Letters* **95** 61-65
[3] Lisin V and Shegeda M 2012 *JETP Letters* **96** 328-332
[4] Abdulsabirov R, Antipin A, Korableva S et al. 1988 *Izv. Vyssh. Uchebn. Zaved., Ser. Fiz.* **2** 24-27
[5] Sattler J and Nemarich 1971 *Phys. Rev.* B **4** 1
[6] Kulpa S 1975 *J. Phys. Chem. Solids* **36** 1317
[7] Macfarlane R M, Cassanho A and Meltzer R S 1992 *Phys. Rev. Lett.* **69** 542

# Ultrafast Below-Band-Gap Laser Pulse Induced Relaxations in CdS Crystal

**A G Shmelev[1], A V Leontyev[1], V G Nikiforov[1], K V Ivanin[1], V S Lobkov[1], O Kh Khasanov[2] and V V Samartsev[1,3]**

[1]Zavoisky Physical-Technical Institute, Sibirsky tract 10/7, 420029, Kazan, Russia
[2]Scientific-Practical Material Research Centre of the Belarus National Academy of Sciences, P. Brovki St. 19, 220072, Minsk, Belarus
[3]Kazan Federal University, Kremlyovskaya St.18, 420008, Kazan, Russia

E-mail: mailscrew@gmail.com

**Abstract.** We report an experimental study of the intra- and interband transitions in bulk CdS crystal induced by a strong below-band-edge femtosecond laser pulse. An additional peak was observed in spectrally resolved four-wave mixing signal shifted to lower energy and positive time delay.

## 1. Introduction

The evolution of high density spatially inhomogeneous carrier distributions following ultrashort pulse excitation (particularly in II-VI semiconductors [3]) is rather complicated. Time-resolved optical spectroscopy measures the third-order susceptibility $\chi^{(3)}$. Since the excitation is detuned below the band edge and linear absorption is negligible, $\chi^{(3)}$ is mainly formed through the two-photon absorption and nonlinear refractive index Kerr effect [6]. Strong two-photon absorption leads to a hot degenerate electron-hole plasma formation. It manifests as an intense band-edge luminescence.

## 2. Carriers recombination and trapping times evaluation via pump-probe experiments

Laser pulses with an energy of 0.4 mJ (pump) and 0.02 mJ (probe), (central wavelength of 790 nm, duration of 50 fs) are being focused to a $\tilde{5}0$ $\mu$m region of the sample. Time interval value $\Delta t$ between those pulses can be set by an optical delay line in the range of 0–300 ps. As the pump pulse passes through the sample a nonequilibrium electron-hole plasma excites through two-photon absorption, and the state of the former can be monitored measuring the intensity of the probe pulse with a certain delay $\Delta t$ (figure 1b).

Figure 2a shows the time dependence of the pump pulse induced absorption while I is fully open. After $\tilde{1}$ ps the absorption turns to transparency increasing with $96 \pm 8$ ps time constant, which corresponds to the carrier trapping process.

Figure 2b shows the results in the same conditions, but with the 0.5 mm diaphragm I opening. The measured intensity change in this case is mainly due to transient Kerr lens effect. After exposure to the pump radiation refractive index of the sample changes along with the effective focal length of F2, so probe beam is now blocked partially by diaphragm I. The detected signal

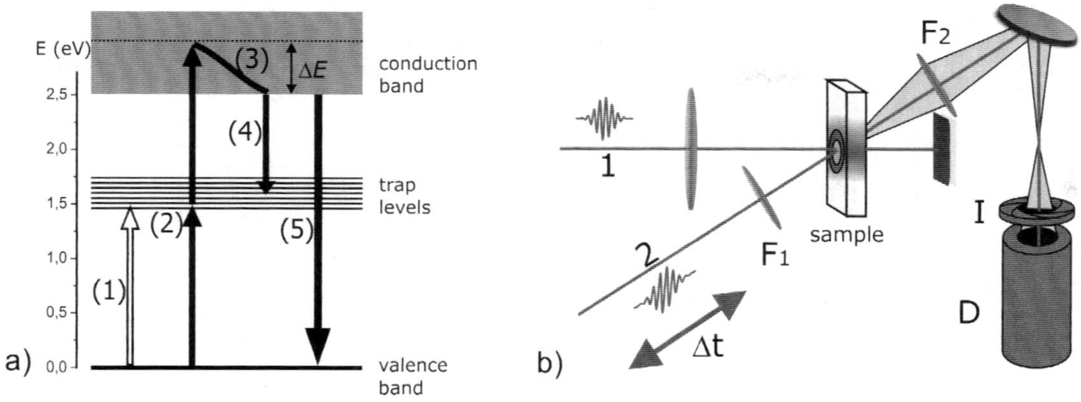


**Figure 1.** (a) – levels and transitions diagram in CdS crystal 1 – probe radiation, 2 - two-photon pump, 3 – hot carriers thermalization, 4 – carrier trapping, 5 – recombination. (b) -– pump-probe experiment setup. 1 – pump pulse, 2 – probe pulse, I -– iris diaphragm, D – PMT (FEU-62).

variation is proportional to the squared transient refractive index, which is determined by free carriers population [2]. Thus, we are able to probe the conduction band depopulation, the measured time constant of the former process is $468 \pm 8$ ps.

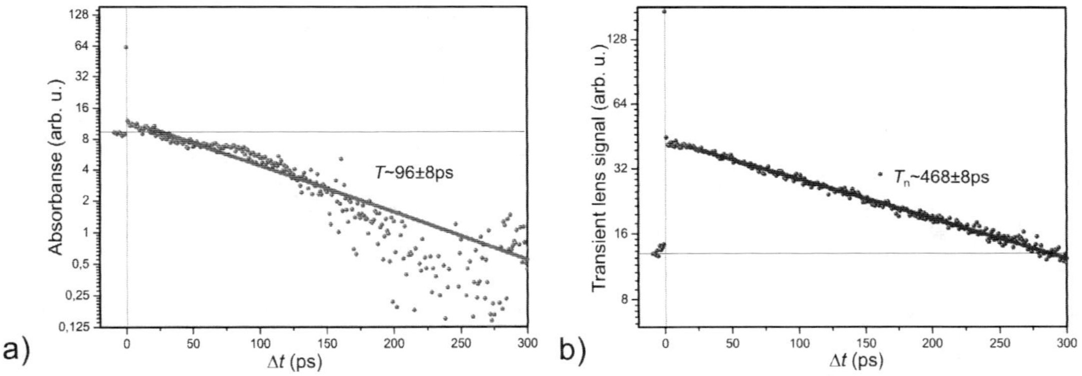


**Figure 2.** The dependence of the induced absorption (a) and transient Kerr defocusing (b) on the probe pulse delay.

## 3. "Fast" transient grating decay. Thermalization slowdown

In this experiment, the pump beam is split into two parallel beams, then focused on the sample and mutually interfered, creating a sinusoidal excited carriers density grating in the sample plane. The grating destruction process is monitored by measuring the diffraction efficiency of the probe beam as a function of delay time $\Delta t$ (figure 3). The transient grating decay appear to have a faster decay component (the time constant of approximately 4.5 ps), which is only present if the energy of exciting pulses exceeds the threshold value of that roughly corresponds to $200GW/cm^2$ intensity. This rapid transient grating decay could be attributed to an ultrafast (350 fs) "explosion" of plasma through stimulated emission recombination of carriers followed by reabsorption of the emitted radiation in regions of lower density [3, 5].

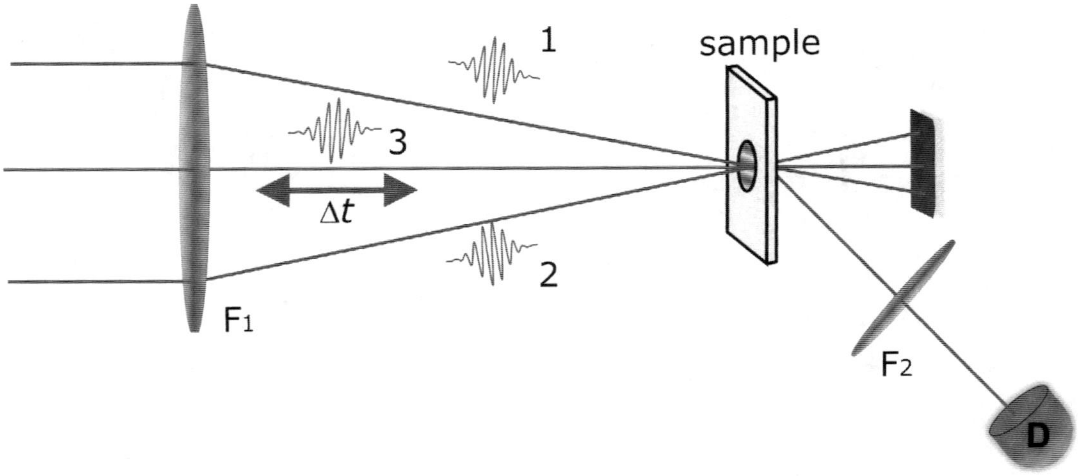


**Figure 3.** transient gratings experiment setup. 1,2 – pump pulses, 3 – probe pulse, D – photodetector.

The hot carriers kinetic energy loss following by its transfer into the crystal lattice is realized mainly through the polar scattering on the LO-phonons with 37.8 MeV energy [4, 5]. The evaluation of the thermalization time is 9 ps, which is much longer than the estimate T ¡400 fs for subthreshold pumping levels [5].

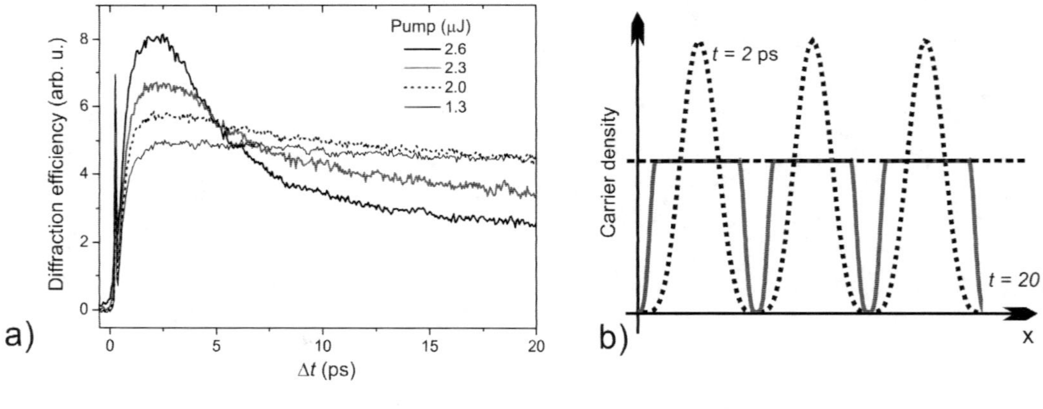


**Figure 4.** (a) – "fast" components in the diffraction efficiency decay. (b) – illustrated "spreading" of the grating when the excited carriers concentration in the antinodes exceeds a threshold.

According to [4], the thermalization slowdown cold be caused by two major reasons: 1) The phonon modes filling, primarily those whose wave vectors correspond to the highest phonon emission rate. 2) Screening of electron(or hole)-phonon interaction, leading to an increase in the longitudinal permittivity of electron-hole plasma.

The energy-time plot of the FWM signal.

The dephasing time Td was studied in a four-wave mixing (FWM) experiment. Spectrally resolved FWM signal, along with normal peak at laser wavelength, showed a weaker peak, offset to lower energy and to positive time delay (figure 5). The energy shift is comparable to the energy of an optical phonon.

## 4. Conclusion
We measured ultrafast transient nonlinear optical properties of CdS at room temperature by femtosecond spectroscopy methods. We associate results from pump-probe experiments with Kerr nonliearity and multi-photon absorption. We find thermalization slowdown at high pump intensities ($200GW/cm^2$) in transient gratings decay experiments. An additional peak was observed in spectrally resolved four-wave mixing signal shifted to lower energy and positive time delay.

## Acknowledgments
This study was supported by Russian Foundation of Basic Research (projects No 14-02-00041-a, No 15-03-02544-a and No 14-02-90000_Bel-a).

## References
[1] Weiner A M 2009 *Ultrafast optics* (John Wiley and Sons)
[2] Brown W E 1993 *Ultrashort processes in condensed matter* (Plenum Press)
[3] Fox E C, van Driel H M 1993 *Phys. Rev. B.* **47** 1663
[4] Dneprovskii V S, Klimov V I, Novikov M G 1990 *JETP Lett.* **52** 534
[5] Lami F, Gilliot P, Hirlimann C 1996 *Phys. Rev. Lett.* **77** 1632
[6] Gur-Arie I and Bar-Ad S 2002 *Phys. Rev. B* **65** 12531

Journal of Physics: Conference Series **613** (2015) 012015 doi:10.1088/1742-6596/613/1/012015

# Simulating single-photon sources based on backward-wave spontaneous parametric down-conversion in a periodically poled KTP waveguide

**A A Shukhin[1], D O Akatiev[2], I Z Latypov[1], A V Shkalikov[1] and A A Kalachev[1,2]**
[1]Zavoisky Physical-Technical Institute, 10/7, Sibirsky tract str., 420029, Kazan, Russia
[2]Kazan Federal University, 18, Kremlevskaya str., 420008, Kazan, Russia

E-mail: spaun73@mail.ru

**Abstract.** The properties of the backward-wave spontaneous parametric down-conversion (SPDC) in a periodically poled potassium titanyl phosphate (KTP) waveguide are studied in the context of creating narrowband heralded sources of single-photon states. The effective index of refraction and spatial profile of different waveguide modes, efficiency of different SPDC processes and purity of heralded photons are calculated numerically for a given waveguide. Compared to the usual co-propagating SPDC, spectral narrowing of the backward-wave SPDC was observed as should be expected.

## 1. Introduction

Developing narrowband sources of single photons and correlated photon pairs is an important part of quantum optics and quantum information [1]. In particular, single-photon wave packets of a sub-GHz or GHz bandwidth can effectively interact with atomic systems and therefore are demanded for implementation of quantum computing and quantum communication protocols involving optical quantum memories [2]. A promising approach to the problem is based on the backward-wave spontaneous parametric down-conversion (SPDC) in a nonlinear waveguide [3-5]. First, as with all SPDC-based sources, the photon number correlation between the down-conversion fields can be exploited to herald the existence of one photon by detection of its partner. In doing so, the vacuum state contribution is automatically removed from the output state while the multiphoton state contribution can be removed by using photon number resolving detectors. Second, emitting correlated photons in opposite directions allows one to reduce both SPDC bandwidth and spectral correlations within each photon-pair [3, 4]. As a result, it is possible to prepare single-photon states of high purity and GHz bandwidth at room temperature. Third, enclosing a nonlinear waveguide in a cavity allows one to realize SPDC in a single cavity mode of MHz bandwidth without additional filtering [5] thereby achieving high efficiency of the source. Finally, such a cavity-enhanced SPDC makes it possible to prepare single-photons of various wave forms by modulating the pump field [6].

In the present work, we calculate the properties of such a promising regime of SPDC in a periodically poled potassium titanyl phosphate (KTP) waveguide and report experimental results demonstrating spectral narrowing of the SPDC field.

## 2. Waveguide model

Our waveguide chip (AdvR Inc.) is fabricated from an X-cut KTP crystal by ion exchange. The area with Rb-diffusion (RTP) has a bigger refractive index than the KTP substrate, forming a channel waveguide under the surface. Along the $-z$-direction, the index profile is usually described as

$$n_j(z) = n_j^{KTP} + \Delta n_j \,\mathrm{erfc}(-z/z_0), \tag{1}$$

where $j = Y, Z$, $\Delta n_j = n_j^{RTP} - n_j^{KTP}$, and the crystal surface is assumed to be located at $z = 0$. Figure 1 presents the microscope images of the waveguide cross section with coupled white light for the channels with different width. The extracted average value of the depth parameter $z_0$ proves to be equal to $\approx 8.9$ μm. The length of the waveguide chip is 24 mm.

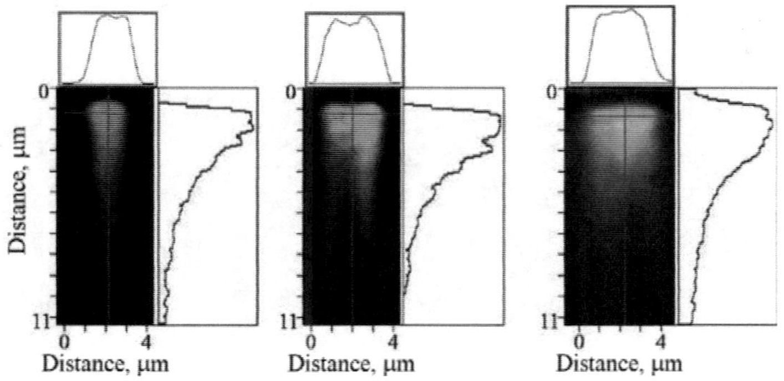


**Figure 1.** Microscope images of the waveguide cross section with coupled white light and corresponding concentration profiles of the Rb dopants (in boxes). The width of the waveguides is 2 μm (left), 3 μm (center) and 4 μm (right).

The effective index of refraction $n_{nm}^{eff}(\omega)$ and spatial profile $u_{mn}(\rho)$, $\rho = (y, z)$ of different waveguide modes were calculated using commercial software package Comsol Multiphysics with RF module. The Sellmeier equations for $n_j^{KTP}(\lambda)$ and $\Delta n_j(\lambda)$ were taken from [7] and [8], respectively. Figure 2*a* shows the simulation region consisting of two rectangles filled with KTP, a rectangular strip of RTP between them and an air region above. The simulation region was chosen to be large enough to make boundary effects negligible. The fundamental modes are shown in figure 2*b*.

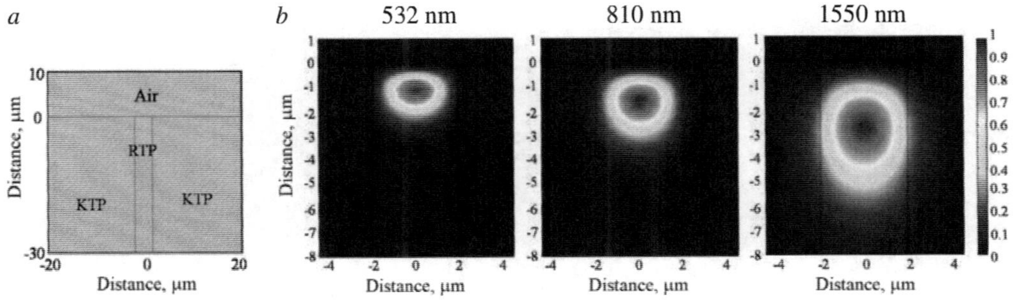


**Figure 2.** The simulation domain (*a*) and intensity profile of the fundamental waveguide modes (*b*) for the pump (532 nm), signal (810 nm) and idler (1550 nm) field polarized along the *Z* axis. The width of the waveguide is equal to 4 μm.

The number of modes supported by the waveguide decreases with increasing wavelength, as should be expected. In particular, there are 16 modes at the wavelength around 532 nm, 6 modes at around 800 nm and only 1 mode at the wavelength 1500 nm. The calculated effective refractive indexes of some low order modes are shown in figure 3.

Journal of Physics: Conference Series **613** (2015) 012015 doi:10.1088/1742-6596/613/1/012015

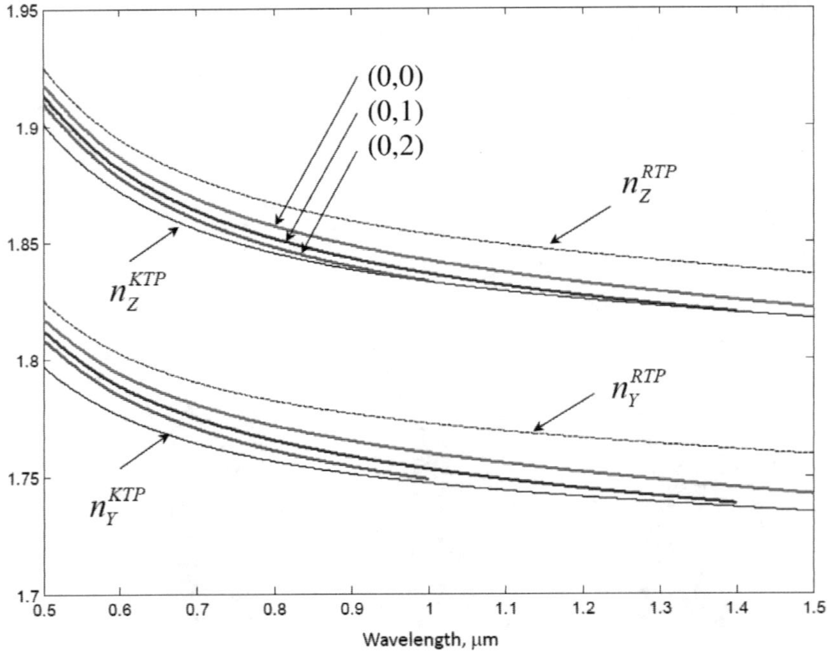


**Figure 3.** The effective index of refraction for some low order modes $(m,n)$ as a function of the free space wavelength for polarization along the crystal $Y$ and $Z$ axes.

## 3. Numerical analysis

The process of SPDC involves sending a pump laser beam into the nonlinear waveguide, where the nonlinear interaction occasionally leads to the annihilation of a high-frequency pump photon and the simultaneous creation of two lower frequency photons, which are usually referred to as signal and idler photons. The photons have the same vertical (horizontal) polarization in type 0 (type I) SPDC and have orthogonal polarization in type II SPDC. Possible variants of SPDC processes are shown in figure 3. The two-photon state of the generated light can be calculated by applying the first-order perturbation theory of quantum mechanics, and the state vector of the generated light can be written as [9]

$$|\psi\rangle = |0\rangle + A\eta_{m_p n_p} \iint d\omega_s d\omega_i \, \eta_{m_p n_p, m_s n_s, m_i n_i} F(\omega_s, \omega_i) a^+_{m_s n_s}(\omega_s) a^+_{m_i n_i}(\omega_i) |0\rangle, \tag{2}$$

where $|0\rangle$ denotes the vacuum state, $a^+_{mn}(\omega)$ is a creation operator corresponding to the waveguide spatial mode $u_{mn}(\rho)$, $\int d^2\rho |u(\rho)|^2 = 1$, and frequency $\omega$,

$$\eta_{m_p n_p, m_s n_s, m_i n_i} = \int d^2\rho \, u_{m_p n_p}(\rho) u^*_{m_s n_s}(\rho) u^*_{m_i n_i}(\rho), \quad d^2\rho = dydz, \tag{3}$$

is the overlap integral between the three interacting fields,

$$F(\omega_s, \omega_i) = E_p(\omega_s + \omega_i) \operatorname{sinc}\left[\Delta k(\omega_s, \omega_i) L/2\right] \exp\left[i\Delta k(\omega_s, \omega_i) L/2\right] \tag{4}$$

is the joint spectral amplitude (JSA), $E_p(\omega)$ is the pump spectral amplitude, and

$$\Delta k(\omega_s, \omega_i) = \frac{\omega_p n^{eff}_{n_p m_p}(\omega_p)}{c} - \frac{\omega_s n^{eff}_{n_s m_s}(\omega_s)}{c} \pm \frac{\omega_i n^{eff}_{n_i m_i}(\omega_i)}{c} - \frac{2\pi m}{\Lambda} \tag{5}$$

is the momentum mismatch taking into account quasiphasematching vector $2\pi/\Lambda$ in the order $m$. The sign "+" before the third term on the right hand side of equation (5) corresponds to the backward propagating idler wave. The factor $\eta_{m_p n_p} = \iint d^2\rho u_{m_p n_p}(\rho) E_{in}(\rho)$ takes into account the overlap between the incident (Gaussian) pump mode $E_{in}(\rho)$ and the corresponding pump mode in the waveguide, and $A$ includes all other constants.

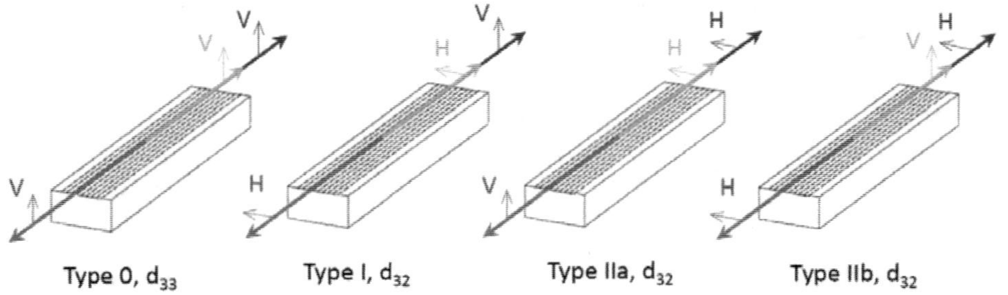


**Figure 3**. Possible types of SPDC in the considered waveguide and relevant nonlinear coefficients. The vertical (V) and horizontal (H) polarizations correspond to the $Z$ and $Y$ axes, respectively.

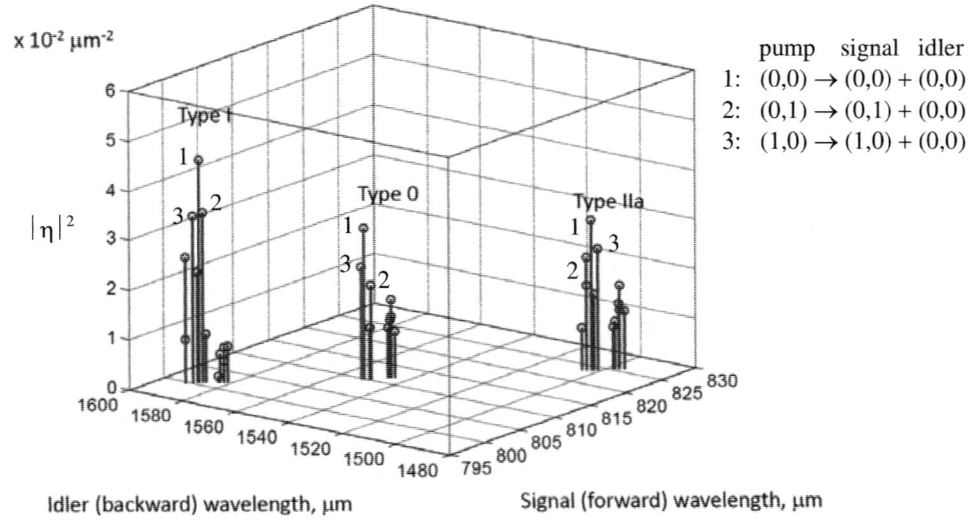


**Figure 4.** Coupling constants $\left|\eta_{m_p n_p, m_s n_s, m_i n_i}\right|^2$ for different SPDC processes in the considered waveguide of 4 µm width. In each case the maximum values correspond to the interaction between the fundamental modes.

The waveguides have an effective poling period of $\Lambda = 2.03\,\mu\text{m}$, which supports SPDC via fifth-order quasi-phase matching ($m = 5$) for 1.5-µm photons emitted in the backward direction. The wavelength of the conjugated photons depends strongly on that of the pump field, which is necessary for maintaining the momentum conservation condition $\Delta k = 0$. In possession of the solutions for the transverse wave equations, we evaluate the interaction efficiency for different mode triplets and different SPDC types, which can be observed under the excitation by 532.2 nm. The most significant variants are shown in figure 4. In a general case, the down-converted fields will be emitted in a superposition of different mode triplets each having a distinct coupling efficiency. However, for a Gaussian pump beam properly aligned with the center of the waveguide most of the energy can be deposited into the (00)-mode and very little is coupled into higher-order waveguide modes. As a result, the process $(0,0) \rightarrow (0,0) + (0,0)$ becomes much more efficient than the other ones.

Regarding the single photon generation, it is important to characterize the spectral or temporal properties of photon pairs through their interference properties. Two photons interfere in a Hong-Ou-Mandel (HOM) interferometer with perfect visibility if they are indistinguishable. In the case where the two interfering photons are heralded from independent sources, indistinguishability leads to the requirement that the JSA is factorable. To be more precise, consider the experimental situation depicted in figure 5a, where two heralded photons corresponding to the same mode of two identical sources interfere at a beam splitter. The four-photon coincidence rate $P_c(\tau)$ as a function of the delay between the two interfering photons $\tau$ is calculated as [10]

$$P_c(\tau)=\frac{1}{2}\left[1-\iint\iint d\omega_1\, d\omega_2\, d\omega_3\, d\omega_4\, F(\omega_1,\omega_2)F(\omega_3,\omega_4)F^*(\omega_1,\omega_4)F^*(\omega_3,\omega_2)e^{i(\omega_1-\omega_3)\tau}\right], \qquad (6)$$

where JSA is assumed to be normalized. Then the HOM visibility $V=1-2P_c(0)$ corresponds to the purity of the heralded photons. In particular, consider type II SPDC process in the fundamental modes (figure 5b) and the pump field of the wavelength 532.2 nm. Figures 5c and 5d present calculated HOM dips which should be observed for the optimal duration of the pump pulse (about 6 ps). The latter is assumed to be Gaussian one. The corresponding purity of the photons is about 0.96. The duration of the single-photon pulse (FWHM) emitted in the backward direction is about 300 ps, which corresponds to the spectral width 2.9 GHz. Similar results are obtained for the other SPDC processes.

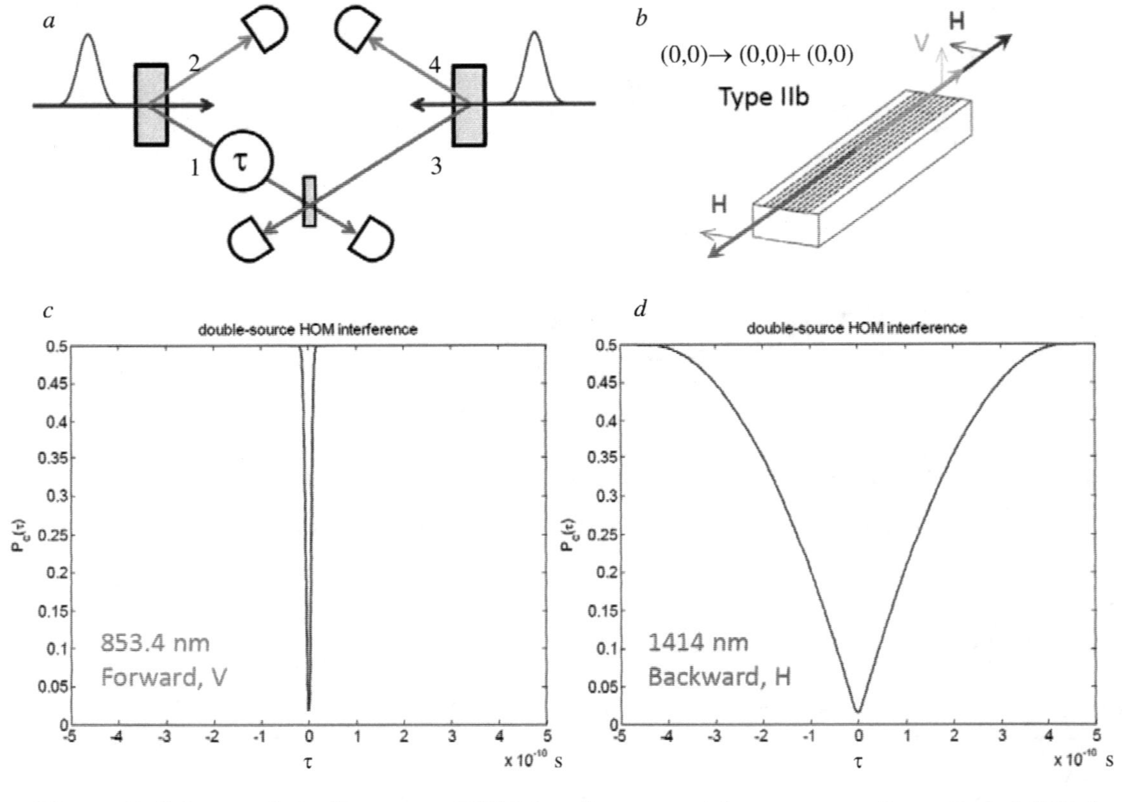


**Figure 5.** Scheme of double-source HOM interferometer (*a*), geometry of considered SPDC process (*b*), and calculated HOM dips for interfering heralded photons emitted in forward (*c*) and backward (*d*) directions. The duration of the Gaussian pump pulse is 6 ps.

## 4. Experimental results

In our experiment we pumped the nonlinear waveguide with the second harmonic of a CW neodymium laser (ALPHALAS MONOPOWER-532-100-SM) and measured the spectrum of the radiation emitted in the forward direction at around 810 nm using a single-photon detector (Perkin Elmer SPCM-AQR-14FC). The pump beam was focused on the waveguide with a 10x microscope objective and the pump beam spot at the entrance of the waveguide was aligned to excite mostly the fundamental pump mode. The emitted radiation, which was collected by a 40x objective, went through an interference filter of 10-nm bandwidth, polarization filter and an air spaced Fabry-Perot resonator (IT 51-30). In doing so we managed to identify a narrow peak corresponding to type-0 SPDC on a strong fluorescence background. The expected bandwidth of the signal field (0.006 nm) proves to be much smaller than that of our Fabry-Perot resonator (0.03 nm), and we were not able to measure the spectrum of the SPDC field directly. However, the observed peak corresponds to the signal field bandwidth not larger than the resonator one. The latter in turn is five times smaller than the bandwidth of usual co-propagating SPDC (~ 0.15 nm) which would be observed under the same conditions. Therefore, we can conclude that significant reduction of SPDC bandwidth was observed.

## 5. Conclusion

We have studied the spectral properties of two-photon states and purity of heralded one-photon states produced by backward-wave spontaneous parametric down-conversion in a periodically poled KTP waveguide, which provides fifth order quasiphasematching for such a nonlinear interaction. Compared to the usual co-propagating SPDC regime in the same material, significant reduction of SPDC bandwidth is observed. According to the numerical simulations, a single-photon source with the purity about 0.95 and bandwidth about 3GHz may be realized on the basis of the investigated material.

## Acknowledgments

This research was supported by the Russian Science Foundation (Grant No. 14-12-00806).

## References

[1] Eisaman M D, Fan J, Migdall A and Polyakov S V 2011 *Rev. Sci. Instrum.* **82** 071101
[2] Sangouard N, Simon C, de Riedmatten H and Gisin N 2011 *Rev. Mod. Phys.* **83** 33
[3] Booth M C, Atatüre M, Di Giuseppe G, Saleh B E A, Sergienko A V and Teich M C 2002 *Phys. Rev. A* **66** 023815
[4] Christ A, Eckstein A, Mosley P J and Silberhorn C 2009 *Optics Express* **17** 3441
[5] Chuu C S and Harris S E 2011 *Phys. Rev. A* **83** 061803(R)
[6] Kalachev A 2010 *Phys. Rev. A* **81** 043809
[7] König F and Wong F N C 2004 *Appl. Phys. Lett.* **84** 1644
[8] Cheng L K, Cheng L T, Galperin J, Hotsenpiller P A M and Bierlein J D 1994 *J. Cryst. Growth* **137** 107
[9] Fiorentino M, Spillane S M, Beausoleil R G, Roberts T D, Battle P and Munro M W 2007 *Optics Express* **15** 7479
[10] U'Ren A B, Banaszek K and Walmsley I A 2003 *Quantum Inf. Comp.* **3** 480

# Investigation of structure of noise component in molecular spectra

**M E Sibgatullin**[1], **D Z Galimullin**[1], **D I Kamalova**[1] **and M Kh Salakhov**[1,2]

[1] Kazan Federal University, Kazan 420008, Russia
[2] Tatarstan Academy of Sciences, 420111, Bauman St, 20, Kazan, Russia

E-mail: sibmans@mail.ru

**Abstract**. In the work the noise component of molecular spectra is investigated by use the Hurst index which is the quantitative parameter of memory effects. It was shown that the low-frequency noise component predominates. The Hurst index is from 0.6 to 0.7 for the studied FTIR spectra registered at different temperatures.

## 1. Introduction

In applied spectroscopy the presence of an experimental noise influences on the mathematical processing of weak spectral absorption bands. An existence of the noise caused by both the nature of the studied object and the recorder device, in the case of the small signal/noise ratio does not allow to restore the spectral contour and, therefore, to estimate the important physical parameters: an intensity, a width, a maximum, a number of components of complex spectral contours. The investigation of the noise structure is an important step of the spectroscopic experiment and it is directed at a reduction of errors to obtain reliable data about a studied object.

For successful and effective use of different mathematical methods of processing of molecular spectra the approaches which allow to take account of a priori information on the nature of useful signal and noise, are of important significant. We would remind you that a noise is usually defined by the recorder noise which does not depend on the signal value in infrared spectral range. The recorders for visible range are much better than ones for infrared range: separate photons can be registered even. For interferometers which are a basis of the Fourier-spectrometers the noise is proportional to a square root of source intensity [1].

In this work, to investigate the noise component structure we used the Hurst index ($H$) which is the quantitative characteristic of memory effects. It characterizes the noise from the point of view of a prevalence of a low-frequency or high-frequency component in this experimental noise. We studied the noise which exists at the IR spectra of 1,2-dichloroethane introduced in branched copolymer of methylmethacrylate as a conformational-inhomogeneous probe to study of local mobility of macromolecules [2]. IR spectra were registered by a FTIR spectrometer Spectrum 100 Perkin Elmer in the range of 400 - 4000 $cm^{-1}$ at the various temperatures from 100 to 300 K. The mathematical treatment of 44 spectra was carried out.

## 2. Separation and analysis of the noise component from molecular spectra

For separation of the noise component of spectra it is necessary to carry out the denoising procedure of the low-frequency component from a signal. Considering that fact that molecular spectra are smooth, it is possible to carry out smoothing procedure with the subsequent removal of the low-frequency component. In this work the addition of the smallest distortions of spectral contours which are characterized by the high signal/noise ratio was used as a criterion of smoothing.

One of the simplest and widespread methods of mathematical processing of the signals distorted by a random noise is the Savitzky-Golay algorithm. The computing algorithm or the filter for smoothing represents an operation of linear weighing of values of entrance data. Weight multipliers are simply filter coefficients. The general procedure of smoothing can be presented by the following form [3, 4]:

$$\varphi_n = \sum_{k=-N_B}^{N_B} a_k f_{n-k}, \tag{1}$$

where $a_k$ are the weight coefficients of the filter, $2N_B$ is a filter width, $\varphi_n$ represents the convolution of an experimental signal $f_n$ with sequence of coefficients $\{a_k\}$. In the Savitzky-Golay algorithm the coefficients of the filter are defined from the condition:

$$\min\|f - Va\|_{L_2}, \tag{2}$$

where $f$ is the vector with length of $n$, $a$ is the vector with length of $m$+1, $V$ is the Vandermonde matrix $(n\times(m+1))$, determined by the equation: $V_{ik} = i^{k-1}$.

In figure 1a one of the experimental spectra is given and the fragment of the spectrum in the range of 800-525 cm$^{-1}$ is shown by the increased scale. As seen from figure 1b, for the absorption band at 750 cm$^{-1}$ the signal/noise ratio is big, and the distortions are shown generally on its contour wings while for the absorption bands at 650 and 710 cm$^{-1}$ the signal/noise ratio is small, and there is a distortion of the spectral contour as a whole. The example of the noise component at the interval of 2075-1775 cm$^{-1}$, for which the Hurst index was calculated, is given in figure 1b.

One of the methods of the optimum estimation of $H$ is the statistical method of rated scope or $R/S$ - analysis which was offered by Hurst [3]. Let $\{x(t)\}$ is a random process. For the analyzed selection $\{x_i = x(t_i)\}(i = 0,...,T)$ a sum of random values is calculated as

$$X_\tau(t) = \sum_{i=1}^{t}(x_i - \langle x\rangle_\tau), \quad 1 \le t \le \tau \le T, \tag{3}$$

where $\langle x\rangle_\tau$ means an average for the selection at the interval $[0,\tau]$. Then, the scope function $R(\tau)$ as the difference between the maximum and the minimum of $X_\tau(t)$ is calculated by

$$R(\tau) = \max_{t\in[1,\tau]} X_\tau(t) - \min_{t\in[1,\tau]} X_\tau(t). \tag{4}$$

Hurst showed empirically that $R/S$ ratio is written by the power law:

$$R/S = (a\tau)^H, \tag{5}$$

where $a$ is any parameter, $H$ is the Hurst index, $S$ is the mean-square distance which is determined by

$$S(\tau) = \left[\frac{1}{\tau}\sum_{i=1}^{\tau}(x_i - \langle x\rangle_\tau)^2\right]^{1/2}. \tag{6}$$

Thus, it is possible to estimate the Hurst index when the experimental data are approximated by equation (5).

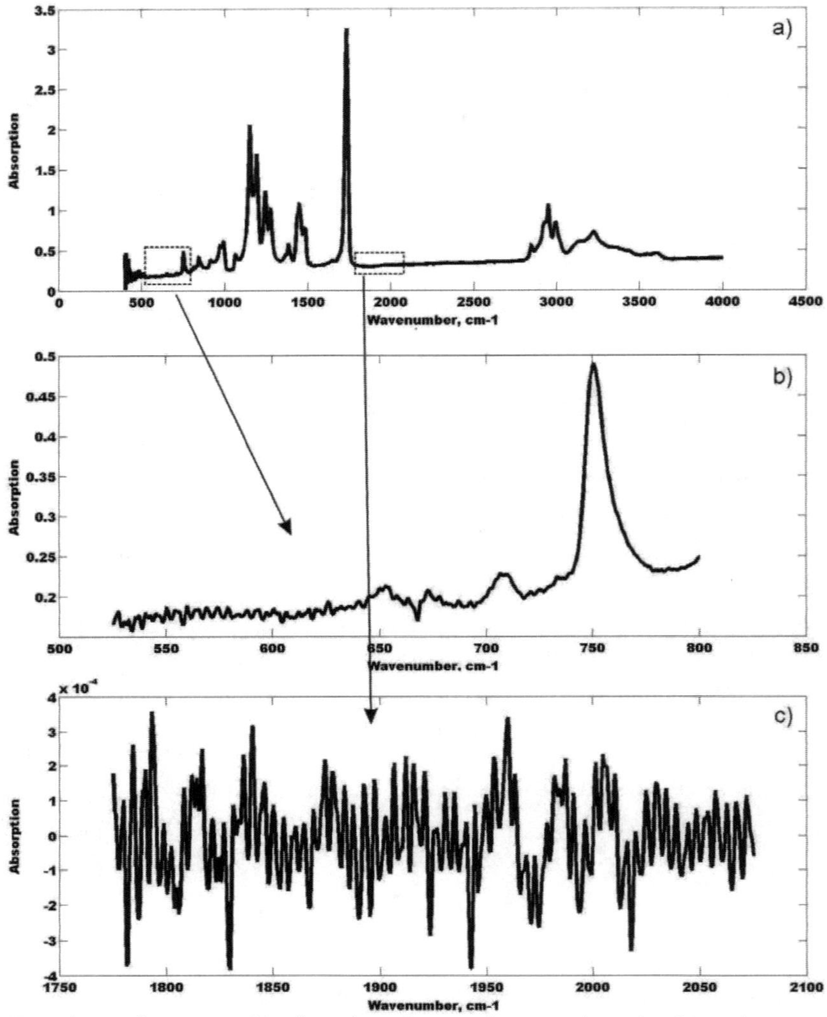


**Figure 1.** Experimental spectrum (a), distortion of spectral contour by noise (b), noise component (c).

To provide a statistical homogeneity of the results of $R/S$ measurements we used the procedure given in [4]. $N$ random sequences (length $\tau$) are selected from the analyzed selection with equal probabilities. For each sequence the relation $R/S$ is calculated. From initial time series at the fixed $\tau$ we obtain $N$ independent $R/S$ values. The results corresponding to the same value of $\tau$, are averaged, and the values $R/S$ are postponed in double logarithmic scale as a function of $\tau$. Further, the obtained values $R/S$ are approximated according to the law (5) by the least-squares method, following which the estimation of the parameter $H$ is carried out. Since the length of the analyzed noise is rather small (600 points), it is necessary to estimate an accuracy of calculations for the determination of a reliability of the obtained results. The calculation error was defined as follows. With the Student distribution the confidential interval for parameter $H$ was defined

$$\left(\overline{H} - t_\gamma \frac{s}{\sqrt{n}}, \overline{H} + t_\gamma \frac{s}{\sqrt{n}}\right), \tag{7}$$

Journal of Physics: Conference Series **613** (2015) 012016 doi:10.1088/1742-6596/613/1/012016

where $\overline{H}$ is an average value of $H$, $t_\gamma$ is the Student coefficient, $n$ is an amount of sampling, $s = \sqrt{\dfrac{\sum_{i=1}^{k} n_i (H_i - \overline{H})^2}{n-1}}$ is a dispersion. The confidence interval covers the $H$ with the reliability $\gamma$. We used the following values of the parameters: $n$=10, $\gamma$ = 99. The maximum error of the calculations was not in excess of 5%.

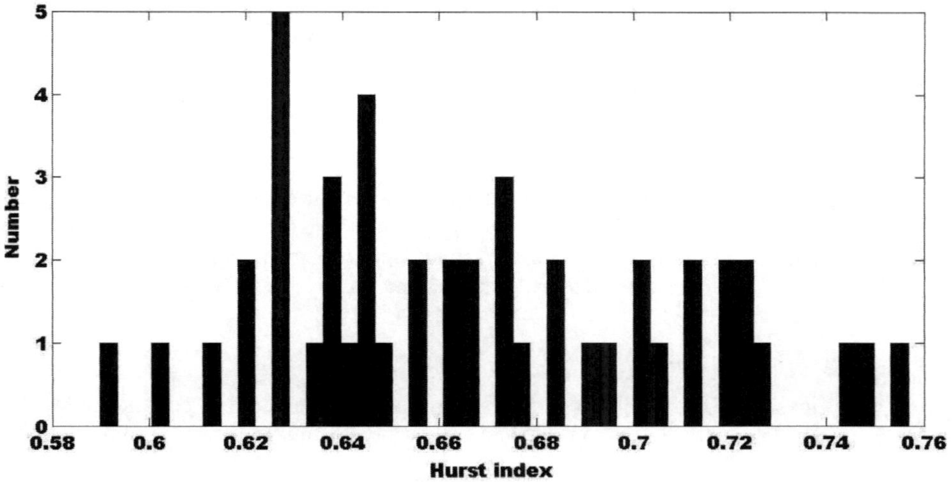

**Figure 2**. Histogram of the Hurst index.

The histogram of the Hurst index calculated for the noise component of the spectra for different temperatures is given in figure 2. We would remind you that noises can be classified formally as follows: the high-frequency noise ($0 < H < 0.5$), uncorrelated (white) noise ($H = 0.5$) and low-frequency noise ($0.5 < H < 1$). The low-frequency noise component becomes predominate when $H > 0.5$. Our investigations showed the presence of the low-frequency noise structure at FTIR spectra of 1,2-dichloroethane introduced in branched copolymer of methylmethacrylate. The Hurst index is in the range from 0.6 to 0.7 generally for all spectra registered at different temperatures. Therefore, when mathematical processing of experimental data it is necessary to take account of the low-frequency noise component, for example, to execute the procedure of the preliminary removal of a noise using the wavelet-analysis with the Hurst index obtained from the analysis of experimental data.

## References

[1] Bell R J 1972 *Introductory Fourier Transform Spectroscopy* (New York and London: Academic Press)
[2] Kamalova D I et all 2008 *Konformatsionnie zondi v izuchenii lokalnoi podvizhnosti polimerov* (Moscow: Fizmatkniga)
[3] Savitsky A 1964 *Anal. Chem.* **36** 1627
[4] Salakhov M Kh and Kharintsev S S 2001 *Matematicheskaya obrabotka i interpretatsiya spektroskopicheskogo eksperimenta* (Kazan, KSU)

Journal of Physics: Conference Series **613** (2015) 012017 doi:10.1088/1742-6596/613/1/012017

# Highly rough tapered gold and silver tips for polarization-controlled TERS performance

**V E Vasilchenko**[1], **S S Kharintsev**[1] **and M Kh Salakhov**[1,2]

[1] Department of Optics and Nanophotonics, Institute of Physics, Kazan Federal University, 420008, Kremlevskaya St, 16, Kazan, Russian Federation
[2] Tatarstan Academy of Sciences, 420111, Bauman St, 20, Kazan, Russian Federation

E-mail: valeriya.vasilchenko@kpfu.ru

**Abstract**. This paper describes an electrochemical method for producing rough silver and gold taper tips for polarization-controlled TERS performance. This method allows to explore the directional intrinsic properties of the sample due to different dipole moment orientations at the tip apex. We give recommendations for fabricating rough silver and gold tips.

## 1. Introduction

Investigation of optical antennas [1,2] properties and parameters resulted in considerable advances of tip-enhanced Raman scattering (TERS) [3-7]. TERS tool utilizes a metallic tip (antenna) that is coupled to the incident light. Due to the resonant excitation of the localized surface plasmon polaritons (LSPPs) at the tip apex, a strongly enhanced and confined light field is created near the tip apex, which provides extremely high spatial resolution in TERS measurements. TERS provides the yield of the spectroscopic information of single objects and structures at nanometer scale and with high detection sensitivity. Tip geometry and material are two of the most important aspects in fabricating tips to generate localized field enhancements. For plasmonic implementations, in particular TERS, silver and gold are the most popular materials. Silver tips have a higher enhancement than gold tips in the visible range, because the imaginary part of its permittivity is much smaller than that of other metals in the visible range [8] and, thus, the damping of the plasmon polaritons is smaller. Unfortunately, the silver probes have such disadvantage as a fast oxidation. One observed that the enhancement of the Ag tips decreased over half a day [9]. The authors relate this decreasing to oxidation.

A new method of polarization-controlled TERS is a powerful tool to investigate directional properties of samples [10]. For performing polarization-controlling TERS one needs tips with nanoparticles with different directions of dipole moments. Polarization-controlled TERS can give information about molecular orientation of molecules at the nanoscale. For example, the tip dipole generates substantially polarization in its oscillation direction, so the SWNTs oriented in the direction of a dipole oscillation were strongly excited [10]. Shapes of random nanoparticles at the tip apex which cause dipole orientation cannot be controlled by existing techniques, but the authors of Ref. 10 offered to measure the direction of the dipole moment of given tip before TERS experiments. This knowledge can provide more information about specimen, because different directions of dipole oscillations can enhance different modes of oscillations of molecular structures. In their study the authors utilized atomic force microscopy (AFM) tips covered by silver. The influence of the dielectric under silver can reduce the optical performance of the coated AFM probes [11] so the solid tips are preferable.

We presented the electrochemical method for preparing tips with rough surface because such structures can help to visualize the samples more precisely. We made solid tapered tips from gold and silver, the most popular materials for plasmonic implementations.

## 2. Results and discussion

Journal of Physics: Conference Series **613** (2015) 012017 doi:10.1088/1742-6596/613/1/012017

For preparing silver and gold tapered tips we used the electrochemical method [12]. In both cases of gold and silver, a wire was positioned in the center of a gold ring counter electrode of 10 mm in diameter made from 2 mm gold wire. A glass beaker utilized in our etching setup was similar to the previous studies [13]. The etching was carried out at the room temperature and the volume of an etchant in the beaker of 160 ml.

A 250 μm Ag wire had been used for preparing silver tips. At our early experiments perchloric acid $HClO_4$ with ethanol had been used as an etchant, but prepared tips were very blunt with apexes radii ~ 5 microns. After few attempts we decided to use nitric acid. The etching solution consisted of $HNO_3$ and distilled water in proportion of 1:1. It is known that nitric acid wet etches Ag as follows:

$$3Ag+4HNO_3 \rightarrow 3(Ag^+[NO_3]^-)+2H_2O+NO,$$

applied voltage accelerates this reaction [11]. The wire was dipped into the solution for the magnitude of 1.5 mm. For precision positioning and pulling out prepared tips we used a nanomanipulator NM3D (DTI, USA). Tips were etched at a dc voltage of 1.5 V, the voltage was supplied with a commercial potentiostat μ-Autolab III (Autolab, The Netherlands). Etching time was in the range of 100-110 seconds. When the etching process comes to an end and the lower part of the wire dropped off, the prepared tip was retrieved from the solution. After that the probes was rinsed with distilled water and explored firstly by optical microscopy and scanning electron microscopy (SEM). SEM images of the three of prepared tips is shown in figure 1. Rough surface structure took place. In the inset we can see a tip apex. The tip is covered by nanoparticles, we suppose that the tips with similar structure can help to yield more information about directional properties of specimen, because of different dipole moment of the nanoparticles on its apex.

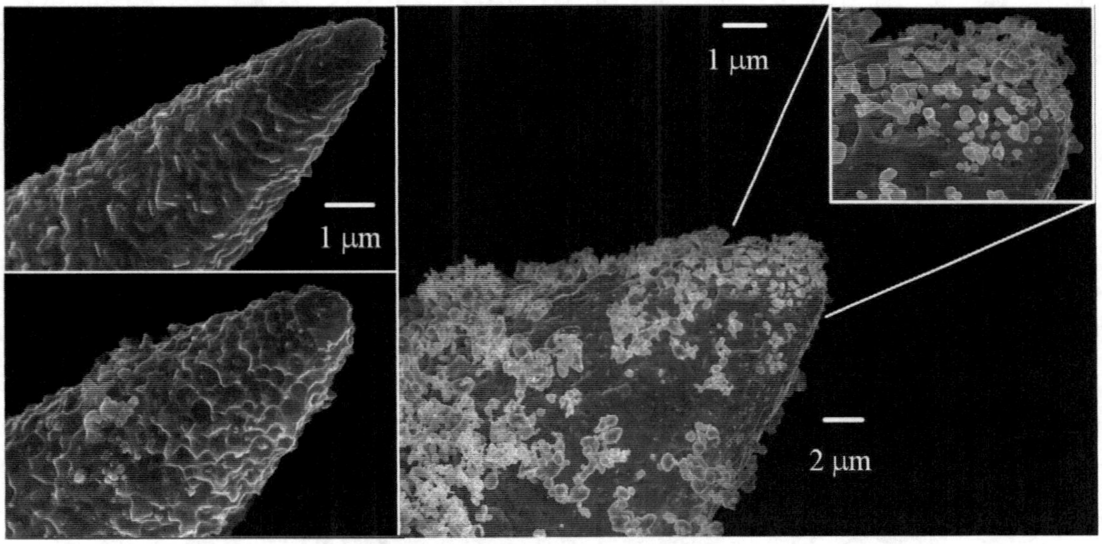


**Figure 1.** SEM image of prepared silver tips.

Gold tips were fabricated from 100 μm wire. The set up is similar to the one for fabrication of silver tips. We chose a hydrochloric acid with ethanol and water as etchant solution. Gold dissolution at hydrochloric acid is a well known process [13,14]. The fastest etching reaction is going in the liquid/solid interface i.e. in the meniscus on the wire. The gold etching occurs due to the following reaction:

$$Au^{3+} + 4Cl^- \rightarrow AuCl^-_4.$$

The complexation occurs due to the reaction between chloride ions with gold 3+ ions. The solution was prepared in following proportions: HCl - 55%, ethanol - 25% and water - 20%. This proportion was chosen experimentally. Ethanol and water were used as solvents. Gold etching drives faster at the meniscus area because of migration of $Cl^-$ ions from solution to the interface. This effect is caused by changing pH balance near the surface [13].

In our study tips were etched with a voltage of 1.7 V, that was taken from the voltamperogramm (figure 2).

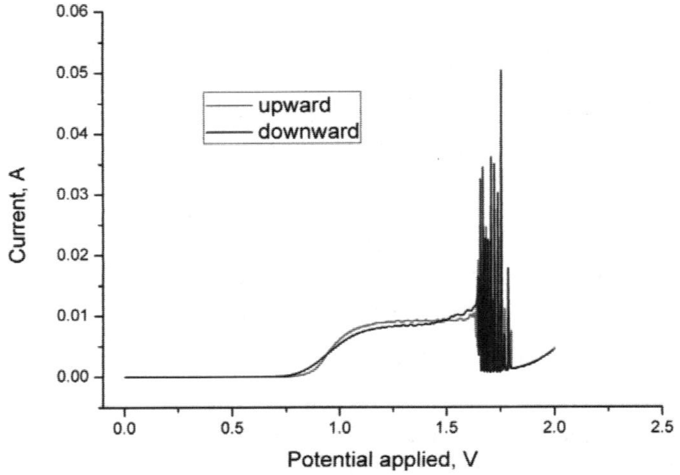


**Figure 2.** Cyclic voltamperometry of electrochemical cell with gold electrodes and etching solution (HCl - 55%, ethanol - 25% and water - 20%).

Figure 2 shows the voltamperogramm of gold etching in the electrolyte. In the previous studies [15] we chose a potential from the range of 1-1.5 V to reduce bubbling and to obtain smooth and sharp tips. A current abruptly drops when a potential exceeds 1.7 V because of blocking an access of $Cl^-$ ions to the wire, as shown in figure 2. An etching time is about 40-50 seconds. The wire was dipped into the solution by a magnitude of 1.5 mm. The shapes of prepared gold tips were explored by transmitting electron microscopy (TEM). As follows from the figure all the tips etched with the above mentioned method are rough.

**Figure 3**. Prepared gold tips and their apexes.

## 3. Concluding remarks

In this paper we have presented methods which allow one to prepare highly rough silver and gold tips for polarization-controlled TERS. For producing rough silver tips the nitric acid/water etchant should be utilized. Appropriate voltage regime of 1.5 V dc voltage was empirically found. For gold tips with high

Journal of Physics: Conference Series **613** (2015) 012017 doi:10.1088/1742-6596/613/1/012017

surface roughness we used hydrochloric acid/ethanol/water etchant and 1.7 V dc voltage. Prepared tips were explored by scanning and transmitting electron microscopy.

## References

[1] Bharadwaj P, Deutsch B and Novotny L 2009 *Advances in Optics and Photonics* **1** 438
[2] Novotny L and Hecht B 2006 *Principles of Nano-optics* (Cambridge: Cambridge University Press)
[3] Kneipp K, Moskovits M and Kneipp H 2006 *Topics Applied Physics* **103** 217
[4] De Angelis F, Das G, Candeloro P, Patrini M, Galli M, Bek A, Lazzarino M, Maksymov I, Liberale C, Andreani L C and Di Fabrizio E 2010 *Nature Nanotechnology* **5** 67
[5] Mauser N and Hartschuh A 2014 *Chem. Soc. Rev.* **43** 1248
[6] Yang Y, Li Z, Nogami M, Tanemurad M and Huanga Zh 2014 *RSC Advances* **4** 4718
[7] Kharintsev S S, Hoffmann G G, Fishman A I and Salakhov M Kh 2013 *Journal of Physics D: Applied Physics* **46** 145501
[8] Johnson P B and Christy R W 1972 *Phys. Rev* B **6** 4370
[9] Zhang W, Yeo B S Schmid and T Zenobi R 2007 *The Journal of Physical Chemistry* C **111** 1733-1738
[10] Mino T, Saito Y and Verma P 2014 *ACS Nano* **8** 10 10187
[11] Lloyd J S, Williams A Rickman R H, McCowen A and Dunstan P R 2011 *Applied Physics Letters* **99** 14 143108
[12] Ren B, Picardi B and Pettinger B 2004 *Rev. Sci. Instrum.* **75** 837
[13] Kharintsev S S, Noskov A I, Hoffmann G G and Loos J 2011 *Nanotechnology* **22** 025202
[14] Busev A I and Ivanov V M 1973 *Analytical Chemistry of Gold* (Moscow: Nauka)
[15] Kharintsev S S, Rogov A M and Kazarian S G 2013 *Rev. Sci. Instrum.* **84** 093106

# AUTHOR INDEX

**Institute of Physics**
Dirac House, Temple Back
Bristol BS1 6BE UK

CURRAN ASSOCIATES INC
proceedings
.com